Archaeological Approaches to Dance Performance

Edited by

Kathryn Soar
Christina Aamodt

BAR International Series 2622
2014

Published in 2016 by
BAR Publishing, Oxford

BAR International Series 2622

Archaeological Approaches to Dance Performance

ISBN 978 1 4073 1257 6

© The editors and contributors severally and the Publisher 2014

The authors' moral rights under the 1988 UK Copyright,
Designs and Patents Act are hereby expressly asserted.

All rights reserved. No part of this work may be copied, reproduced, stored,
sold, distributed, scanned, saved in any form of digital format or transmitted
in any form digitally, without the written permission of the Publisher.

BAR Publishing is the trading name of British Archaeological Reports (Oxford) Ltd.
British Archaeological Reports was first incorporated in 1974 to publish the BAR
Series, International and British. In 1992 Hadrian Books Ltd became part of the BAR
group. This volume was originally published by Archaeopress in conjunction with
British Archaeological Reports (Oxford) Ltd / Hadrian Books Ltd, the Series principal
publisher, in 2014. This present volume is published by BAR Publishing, 2016.

Printed in England

BAR titles are available from:

          BAR Publishing
          122 Banbury Rd, Oxford, OX2 7BP, UK
EMAIL   info@barpublishing.com
PHONE +44 (0)1865 310431
  FAX   +44 (0)1865 316916
          www.barpublishing.com

TABLE OF CONTENTS

Archaeological Approaches to Dance Performance: An Introduction   1
    Kathryn Soar and Christina Aamodt

Archaeology of Dance   5
    Yosef Garfinkel

Who is the Ancient Egyptian Dancer?   15
    Batyah Shachter

Dance Dating in the Old Kingdom; Formal Rules, Step 1:
    Know thy Dances. Establishing a Typology of Old Kingdom Dance   31
    Lesley J. Kinney

Dance in the Prehistoric Aegean   47
    Christina Aamodt

The Dance of Death: Dancing in Athenian Funerary Rituals   59
    Hugh Thomas

Dancers' Representations and the Function of Dance in Han Dynasty
    (202 BC – 220 AD) Chinese Society   67
    Marta Zuchowska

Warrior Dance, Social Ordering and the Process of Polis Formation in Early Iron Age Crete   75
    Anna Lucia D'Agata

Revel without a Cause? Dance, Performance and Greek Vase-Painting   85
    Tyler Jo Smith

Archaeological Approaches to Dance Performance: An Introduction

Kathryn Soar and Christina Aamodt

The present volume is the outcome of a session held at the 15th European Archaeological Association conference at Lake Garda in Italy, in September 2009. The theme of this session developed from the editors' research interests in ritual (Aamodt 2006) and performance (Soar 2009) in archaeology, and was organised specifically as an opportunity for archaeologists interested in the use and perception of ancient dance to present and discuss their work in an open forum.

The aim of the session was to consider a topic that was hitherto underexplored in archaeological literature (with notable exceptions, such as Garfinkel (2003)) – the archaeological importance of dance. There are many reasons why dance can be of interest and importance for archaeologists. One of the strongest arguments for turning to dance as a tool and subject for archaeological investigation is its universality. Dance in its basic form, as organised rhythmical physical movement, has been found in all cultures; from engraved ivories at the Upper Palaeolithic of the Geissenklösterle site in Southern Germany (Garfinkel 2010, 208) or the miniature sculpture from Galgenberg, Austria (Bahn 1989) to the vase paintings of Classical Greece and Rome, dance holds a central place in the artistic representation and physical action of cultures spread widely both geographically and temporally.

Archaeology and dance may at first sight seem to be an unlikely pairing, but in fact the archaeology of dance draws on several key themes in contemporary archaeological theory – such as embodiment and phenomenology – which allow a more developed examination of living bodies and human beings, who have emotions, think with their bodies, and engage in actions and performance (Lopez y Royo 2002, 146). These approaches allow for more 'fleshed out' understandings of ancient behaviour and their possible implications (McGowan 2006, 33). The human body is the central focus of archaeological investigation – how does the body experience and shape the world around it? The body here is the medium through which people act, understand their own identities and communicate it to others. In the past, human beings thought with their bodies and engaged in action with their bodies. This is also what happens in dance – people use their bodies to move, they dress them up, they use them to communicate. If archaeology is the study of human activity in the past, then dance is very much part of the scope of human experience. In this regard, dance is as integral for the study of the human past as hunting, eating or any other form of behaviour (Lopez y Royo 2002, 144).

By bringing together the fields of dance and archaeology, it is the aim of this volume to try and make sense of some of the material remains of dance. This is no straightforward matter, however, although these papers can help us understand how to approach some of these key issues. In this regard, Yosef Garfinkel's keynote paper, '*Archaeology of Dance*' presents not only an examination of previous research undertaken on the Archaeology of Dance, but suggests where this sub-discipline can go in the future. Using examples from Europe and the Near East from the Upper Palaeolithic to the Neolithic, he proposes several criteria which can help archaeologists discern evidence for dance in the material record.

When collecting these papers together for publishing, two separate but equally as important strands of analysis became clear – these may be termed 'choreographical perspectives', and 'social perspectives'. The papers are organised around these two approaches. These themes parallel the approach taken by dance scholar Anca Giurchescu: the first approach considers dance as a living phenomenon; the second situates it on the paradigmatic level, comprising the ideological, socio-political, economic and cultural systems which function in a given community (Giurchescu 2001, 109).

Choreographical perspectives

Dance is a common phenomenon but one with a myriad of different meanings, and these differing meanings and functions are represented within the papers found in this volume. The importance for studying dance is that it is a multi-faceted phenomenon, which has an 'invisible' underlying system aside from the outwardly visual (Kaeppler 2000, 117). It is a communicative vehicle, consisting of the visual, kinaesthetic, and aesthetic aspects of human movement (Kaeppler 1992, 196). But basic to all definitions of dance is the concept of rhythmic or patterned movement. In this regard, dance is a rhythmic action done for some purpose transcending utility (Royce 1977, 5).

Within anthropology, dance is viewed as a positive social activity, and anthropologists attempt to discern its function and significance. Dance is seen not as entertainment or as a mode of self-expression, but as "an active force in shaping ideas and social life" (Blacking 1986, 10). However, the nature of the archaeological evidence and the basis of an archaeological investigation into dance have one crucial difference to anthropological studies – there is no dance left to study, per se. For the anthropologist, a key point of study is the ability to go into the field and examine and investigate the dance first hand. This is obviously an impossible task for the archaeologist, whose only clues to the nature of the ancient dance lie in incomplete visual records.

As a result, studying dance archaeologically can raise issues even at this most fundamental of levels; when studying the residues of dance rather than the act itself, how can we tell if something even *is* dance? What are the issues regarding the transformation of a transient and dynamic act into a static depiction? In order to be understood as dance, movements must be grammatical, they must be intended as dance and interpreted as dance (Kaeppler 2000, 118). Batyah Schachter, in her paper entitled '*Who is the Ancient Egyptian Dancer?*' uses the case-study of Egyptian representations to understand and answer some of these questions. In order to do so, Egyptian artistic rules and regulations are considered so that a definition of dance can be drawn up, and from there notions of 'dance' and 'dancer' can be discerned and considered.

Moving on from this preliminary but crucial question, how can we infer from these frozen depictions what kind of movement and rhythm these depictions show? What separates it from other forms of movement? What form does the dance take? And if the form of dance is vital in its reconstruction, is it possible to actually reconstruct it from the only partial remains we have of these acts? Lesley Kinney's paper '*Dance Dating in the Old Kingdom; Formal Rules, Step 1: Know thy Dances*', which also discusses Egyptian dance, takes as its focus the creation of typologies of dance genres via an examination of iconographic representations of dance from the Old Kingdom. By analysing both the representations of dance in art and previous work on the issue, she develops a typology of the various genres of dance which took place during the Old Kingdom period by their form and steps. Her paper not only reconstructs and categorises the varying forms dance took, but also offers a new method for dating tombs in the Old Kingdom.

These two papers – Schachter's and Kinney's – represent one of the two key approaches undertaken when studying the Archaeology of Dance – the choreographical approach, which considers dance from the perspective of the dancer. These two papers focus on the dance itself and the reconstruction of its movement, questioning and analysing the experience of the dancer in the past.

Social perspectives

A second approach which emerged during this session is reflected in the next collection of papers. These papers differ in focus, geography and time period, but all address the socio-political meaning of dance, and consider questions of what dance can do for society and what we as archaeologists can reconstruct about society based on that knowledge. Rather than a choreographically-based perspective, these papers follow an archaeologically-based perspective, one that considers dance in its social context.

The varying functions of dance

Part of the importance of dance is that it affects life from several different perspectives. It is a physical act, an action that is encoded into bodily memory. In this sense, memory is incorporated within the body itself, in bodily postures, activities, techniques, and gestures, learned through cultural practices (Connerton 1989). This physicality of action leads into the cultural aspects of dance, where peoples' beliefs, values, and ideologies shape the style and production of the dance, and as a result reflexively comment on prevalent systems of thought (Hanna 1987, 3). Beyond the physicality of dance – the movement and energy expenditure which is the fundamental of dance - it also encapsulates various behaviours and meanings which are often culturally specific.

Ritual function of dance

One of these behaviours is ritual. Ritual has always had a strongly performative element, as, at the most basic level, it involves 'doing things', but also in that it is often staged and aims to afford participants an intense experience (e.g. Tambiah 1979). Dance is often one aspect of this performativity.

The characteristics of dance make it attractive to religious behaviour. The act of dancing transforms the person, taking him/her out of the ordinary world and placing him/her in a world of heightened sensitivity (Boas 1972, 21). A common theme in the use of dance to facilitate religious experience is the recognition of the power of dance to express, communicate and facilitate certain emotions and states of mind; the power of dance in religious practice lies in its multisensory, emotional, and symbolic capacity to create moods and a sense of situation in attention-riveting patterns by framing, prolonging, or discontinuing communication (Hanna 1987, 203.) This can make dance one of the most effective conveyers of meaning, establishing as it does subliminal communication more effectively than other human social activities (Royce 1977, 196).

Intense, vigorous dancing can lead to an altered state of conscious-ness through brain wave frequency, adrenalin, and blood sugar changes (Hanna 1988, 285). This aspect of the bodily movement has the ability to alter states of consciousness and propel the mind into various stages of trance (Garfinkel 2003, 88). The induction of an altered state of consciousness can have many forms and functions – these include healing rituals, rites of passage, or interaction with the divine.

Through movement, dance can be used to invoke deities or spirits, as Christina Aamodt explores in her paper '*Dance in the Prehistoric Aegean*'. Using both iconographical and architectural evidence, she examines and explores how dance is used for various religious functions – such as epiphany, rites of passage, and funerary ritual – in the communities of mainland Greece, Crete and the Cyclades during the Bronze Age.

Hugh Thomas, in his paper '*The Dance of Death: Dancing in Athenian Funerary Rituals*', combines iconographic with written evidence to examine the performance of dancing as part of Athenian funerary rites from the 8th to the 6th century BC. He concludes that dancing was most likely performed both in relation to the *ekphora* of the deceased and in post-funerary rituals, even though the motif is rarely depicted in comparison to the rite of *prothesis*, the laying out of the deceased. This, according to Thomas, may be due to the importance of the *prothesis*, the viewing of the dead, but may also simply suggest that the Athenians did not consider it necessary to depict the various stages of the funerary rituals.

But ritual dances are not static – they evolve over time and often find new outlets in other forms. Marta Zuchowska's paper, '*Dancers' representations and the function of dance in Han dynasty (202 BC – 220 AD) Chinese society*', explores the ritual origins of certain dances in pre-Tang China, which later developed into a form of state entertainment. Zuchowska argues that clay figurines and jade plaques representing dancers found in the tombs of Han Dynasty tombs have their origins in earlier ritual activities, but by the period of their deposition, the dances became secularized, holding symbolic and ceremonious rather than religious and ritual functions.

These papers illustrate the potential of dance analysis and study to further understand the performative function of ritual and to develop more nuanced readings of cult activity in the past, by interpreting the codes and conventions of ritual dance.

Educative role of dance

Another one of the behaviours manifest through dance is its educative role. Dance has occupied a central place in educational theory since the times of ancient philosophers such as Plato and Aristotle, who believed that dance contributed to aesthetic, moral, and intellectual values as well as to enhancing physical adeptness and overall well-being (Carter 1984, 293). Human dance, in contrast with the dancelike movement of other animals, occurs through mental cultural maps that guide who does what, when, where, how, and why (Hanna 1983, 223). Structured movement systems are systems of knowledge-the products of action and interaction as well as processes through which action and interaction take place- and are usually part of a larger activity or activity system. These systems of knowledge are socially and culturally constructed created by, known, and agreed upon by a group of people and primarily preserved in memory (Kaeppler 2000, 117). In periods before schools and writing, community rituals, symbolised by dance, were the basic mechanism for conveying education and knowledge to the adult members of the community and from one generation to the next (Garfinkel 1998, 231).

These ideas are explored the paper '*Warrior dance, social ordering and the process of polis formation in Early Iron Age Crete*' by Anna Lucia D'Agata, which examines the function of dance in preparing and socialising children for adult roles, in this case the initiation of young males into warriors. Iconographic evidence of dance on a krater from the Protogeometric site of Thronos Kephala on Crete is used to argue that the warrior dance was part of a Cretan education system which marked the transition from boys to men.

Social role of dance

Dance also plays an important social role; in dance, patterns of social organisation, such as the relationships between individuals in groups, and among groups, are both reflected and influenced, and therefore can act as a vehicle both for the articulation of attitudes and values, as well as for control, adjudication, and change (Hanna 1987, 3) Through the study of dance, it is possible to see and understand how identity is marked, shaped, and introduced. Styles of dance form styles of social relations – thus a study of the dance itself and its context can help us establish the underlying ideologies attached to bodily movement and discourse (Soar 2010, 151).

This is of importance to archaeologists, as the structured content of dance can be a visual manifestation of social relations and can help in understanding the structure and values of a society (Kaeppler 2000, 117).

Aamodt also explores this in her paper, arguing that dance performances could function as a means of social control during the later Bronze Age, helping to establish and maintain new social groupings.

Dance is also a valuable tool in deciphering the way in which social relations are 'signaled, formed and negotiated' through bodily movement (Desmond 1993, 34). One of these relations is gender identity- dance is an important means by which cultural ideologies of gender are reproduced (Reed 1998, 516). D'Agata explores this issue in her paper, investigating the construction of male identity through dance performance. This creation of male identity is part of a process of larger group identity; in this case the community's ruling group, whose emergence during this period can be considered as an important stage in the development of a secondary state formation.

Combining approaches

Finally, Tyler Jo Smith's paper combines elements of both of these two approaches to archaeological evidence for dance. In '*Revel without a Cause? Dance, Performance and Greek Vase-Painting*', she examines the importance and efficacy of databases and online collections of Greek vases, specifically the Beazley Archive Database and the *Corpus Vasorum Antiquorum*, for the scholar of ancient dance, based as they are on forms of movement. From there, she examines a specific type of evidence from the databases, Athenian red-figure dance-scenes, mainly decorating drinking-cups, for what they can show us about ancient Greek dance as a whole.

Conclusion

These are only some of the issues and areas which an archaeological approach to dance performance can help to address. This volume will hopefully aid us in our attempts to more fully understand ancient dance and the societies which created them. With such a diverse array of subject matter – from Palaeolithic Europe to Tang Dynasty China – we can see that these issues permeate time and space, and that dance is a language which transcends temporality and spatiality; dance is, and can be, a universal human language.

References

Aamodt, C. 2006. *Priests and Priestesses in the Mycenaean Period*. Unpublished PhD dissertation, University of Nottingham.

Bahn, P. 1989. Age and the Female Form. *Nature* 342, 345-346.

Blacking, J. 1986. Dance as a cultural system and human capability: an anthropological perspective. In Janet Adshead (ed.) *Dance - A Multicultural Perspective. Report of the Third Study of Dance Conference University of Surrey, 5-9 April 1984*. 2nd edition, 4-21. University of Surrey, National Resource Centre for Dance.

Boas, F. 1972. Origins of Dance, In *Dance Therapy: Roots and Extensions. Proceedings of the Sixth Annual Conference of the American Dance Therapy Association*, 21-22. Columbia, American Dance Therapy Association.

Carter, C. L. 1984. The State of Dance in Education: Past and Present. *Theory into Practice* 23(4), 293-299.

Connerton, P. 1989. *How Societies Remember*. Cambridge, Cambridge University Press.

Desmond, J. 1992. Embodying Difference: Issues in Dance and Cultural Critique. *Cultural Critique* 26, 33-63.

Garfinkel, Y. 2010. Dance in Prehistoric Europe. *Documenta Prehistorica* 38, 205-214.

Garfinkel, Y. 2003. *Dancing at the Dawn of Agriculture*. Austin, University of Texas Press.

Garfinkel, Y. 1998. Dancing and the Beginning of Art Scenes in the Early Village Communities of the Near East and Southeast Europe. *Cambridge Archaeological Journal* 8(2), 207-237.

Giurchescu, A. 2001. The Power of Dance and its Social and Political Uses, *Yearbook for Traditional Music* 33. 109-121.

Hanna, J.L. 1988. The Representation and Reality of Religion in Dance. *Journal of the American Academy of Religion* 56(2), 281-306.

Hanna, J.L. 1987. Dance and Religion. In M. Eliade (ed.) *Encyclopedia of Religion*, vol. 4, 203-213. New York, MacMillan.

Hanna, J.L. 1983. Dance and the Child. *Current Anthropology* 24(2), 222-224.

Kaeppler, A. L. 2000. Dance Ethnology and the Anthropology of Dance. *Dance Research Journal* 32(1), 116-125.

Kaeppler, A. L. 1992. Dance. In Richard Bauman (ed.) *Folklore, Cultural Performances and Popular Entertainments: A Communications-Centred Handbook*, 196-203. Oxford, Oxford University Press

Lopez y Royo, A. 2002. Archaeology for dance: an approach to dance education. *Research in Dance Education* 3(2), 143-153.

McGowan, E. 2006. Experiencing and Experimenting with Embodied Archaeology: Re-Embodying the Sacred Gestures of Neopalatial Minoan Crete. *Archaeological Review from Cambridge* 21(2), 32-57.

Reed, S. A. 1998. The Politics and Poetics of Dance. *Annual Review of Anthropology* 27, 503-532.

Royce, A. Peterson. 1977. *The Anthropology of Dance*. Bloomington, Indiana University Press.

Soar, K. 2010. Circular Dance Performances in the Prehistoric Aegean. In A. Chaniotis, S. Leopold, H. Schulze, E. Venbrux, T. Quartier, J. Wojtkowiak, J. Weinhold and G. Samuel (eds.) *Ritual Dynamics and the Science of Ritual: Body, Performance and Agency*, 137-157. Wiesbaden, Harrassowitz Verlag.

Soar, K. 2009. *The Archaeology of Minoan Performance*. Unpublished PhD dissertation, University of Nottingham.

Tambiah, S. J. 1979. A Performative Approach to Ritual. *Proceedings of the British Academy* 65, 113-169.

ARCHAEOLOGY OF DANCE

Yosef Garfinkel
Hebrew University of Jerusalem

Abstract: Dance activity does not leave many tangible remains and therefore has been neglected by archaeologists. However, in this paper, we will discuss how dancing activities can be investigated from an archaeological point of view. The main points to look for are the context of dancing scenes, and dance accessories.

Introduction

Dance is a rhythmical movement which can be classified as a form of non-verbal communication. Dance is not limited to the human species and is performed by various animals such as bees, birds and mammals. In this context it is always performed by a solo individual. In human society, dance is usually performed by groups of people, and in a variety of situations (for general introductions, see, for example, Sachs 1952; Lange 1976; Royce 1977; Hanna 1987; McNeill 1995). In traditional societies dance is a major social activity, as demonstrated by many observations of the dancing activities of the San Bushmen of South Africa (Marshall 1969; Biesele 1978; Katz 1982). Part of this rich ethnographic data, as well as of other human groups, was summarized in *Dancing at the Dawn of Agriculture* (Garfinkel 2003). An important observation is that after hours of rhythmical circular dancing a few of the participants often fell into a trance. The trance was understood to be a form of contact between the community and supernatural powers - in other words a mystical event, the core of religious experience. The clear connection between dance and trance is probably the main reason for the depiction of intense dancing in many religious ceremonies.

Ancient human dancing is a neglected topic in research. Seldom can one find articles dealing with dance, while books are almost non-existent. Most previous research on dance is based on the iconographic analysis of dancing scenes, which is mainly done as art history. Occasionally written accounts were also used, but these should be classified as history. Thus, one may rightly wonder if dance can be investigated at all with archaeological tools, since from an archaeological point of view it is a most elusive human activity.

The study of dance by archaeologists is challenging for two main reasons:

1. Dancing activity does not leave visible remains, so the chances of finding foot-prints in a circle, or a group of human skeletons suddenly trapped and buried during a dance, are minimal. Until relevant data become available, we are dealing with a very fragmented record.

2. Modern archaeological and anthropological research evolved in western civilization, which is dominated by a Christian point of view. Unlike most other human religions, Christianity has a particularly negative approach to dance. In the New Testament the term is mentioned only once, in the extremely dramatic dance of Salome which concluded with the beheading of John the Baptist (Mark 6, 21–26). In contrast, the Old Testament described dancing dozens of times using ten different verbs (Gruber 1981). Indeed, dance is not part of any official Christian liturgy. The unawareness of western scholarship of the importance of dance in human activity must be seen against this background.

This combination of unawareness on the one hand and a very fragmented record on the other placed dance beyond our knowledge. So the first stage in developing an Archaeology of Dance is to create an intellectual environment which recognizes dance as an important human activity, and thus open our minds to the evidence available to reconstruct dance in the past.

In the past I summarized the implications of various ethnographic observations for the study of dance (Garfinkel 2003), some of which have direct implications for the Archaeology of Dance:

1. Dancing is an activity done at the community level and reflects interaction between people.
2. Dancing is performed in an open space, and not within any structure.
3. The activity involves men and women in close proximity, although they do not mix in the same row or circle.
4. Dancing is often performed with special decorative elements: coiffure, head coverings, masks, body paintings and dress. In many cases the dancers use very elaborate accessories whose preparation begins months before the event itself.
5. Dancing is usually performed at night.
6. Dancing is accompanied by rhythmic music: singing,

clapping hands or musical instruments such as drums or rattles.
7. Dance is an ecstatic event which is considered a deep spiritual experience by the participants.

From an archaeological point of view, one would expect a dancing ground to be found outdoors, in large open areas, either within or outside the settlement. A large, open courtyard of a temple, or an open area near a temple would be a preferable location in an urban setting. We could expect a large fire area in the centre, much larger than the average cooking hearth. Musical instruments would be found (but only if they were broken during the ceremony), as well as various dance accessories, such as masks and beads. The ceremonies probably included eating and drinking, so special serving vessels could be found as well. Unfortunately, until recently no such recognized dancing ground has been identified by archaeologists. Below I will propose possible dancing grounds near temples wherein evidence for both dancing and music were found.

Archaeology and Dance

Various aspects of dance can be identified archaeologically.

The dancing scene as an archaeological object

Ancient dancing scenes, which were depicted on stone, pottery, or any other material, are archaeological objects. Thus we have the basic responsibility of establishing their accurate dating, recording their context in the excavated site, ensuring their conservation, and carrying out proper documentation and publication. While these steps would seem obvious and basic requirements, I was quite surprised during many years of research to find dozens of dancing scenes in museum basements, and even on exhibit, which were not published.

Sites rich with depictions of dance

Dancing scenes are not distributed equally between sites. While most archaeological sites did not produce even one dancing scene, there are a few sites with a rich representation of dancing figures. There is a need to give special attention to these sites and to try to clarify why, at this specific place at that specific period, many dancing scenes were produced. While this seems to be a very simple matter, many publication reports of such sites have in fact neglected the dancing aspect and usually no explanation was given as to why that specific site revealed such rich representations of dance activity. A few examples of such sites are listed here:

Gönnerdorf

An open air site on the eastern bank of the Rhine near Koblenz in Germany, it is dated to the Magdalenian culture, c. 14–12 millennium BCE (Bosinski 1970; Bosinski and Fischer 1974). The rich artistic assemblage is composed of 224 anthropomorphic figures engraved on 87 stone plaques and 11 anthropomorphic figurines. The engravings on stone plaques usually represent groups of figures, and isolated representations occur only on broken plaques, so they are probably a part of a larger group. Usually the figures were depicted in a row, one after another, in profile, most often facing to the right, with up to ten such figures in the row (**fig. 1:1**). Another type of engraving depicts only two figures in each scene, facing each other (**fig. 1:2**). All the engravings are representations of girls or young woman in half crouching positions, sometimes with the arms partly raised. The excavators suggested that these figures are dancing (Bosinski 1970, 93–94; Bosinski and Fischer 1974). Indeed, these groups of figures are not presented in daily-life activities, such as hunting, fighting, or holding a baby. The female figures are presented in rows, posed in a dynamic body gesture - both features of dancing.

Dance research commonly classifies dance into three basic types: circle dance, line dance and couple dance (Garfinkel 2003, 41–43). It would appear that at Gönnerdorf we can see two different types of dance (Garfinkel 2010). The figures presented one behind the other in profile may indicate a line or a circle dance. But as they are usually depicted facing to the right, what we probably have here is a circle dance with counter-clockwise movement, typical of dancers in a circle (Garfinkel 2003, 44–47). The scenes with only two figures facing each other probably indicate a couples' dance.

Eleven figurines, c. 5% of the human representations, which depict young females in the same gesture as shown on the engravings were found. Sometimes a few such pendants were found in a pit, indicating that they were meant to represent groups of young females dancing together.

The dancing characteristic of these scenes and figurines is achieved by a several aspects:

a. The same body gestures are repeated for all the individuals.
b. In a row of figures, all the individuals face the same direction of movement.
c. Most of the rows are moving to the right, which in a circle would create a counter clockwise movement.
d. Heads were not portrayed, as the scenes emphasize the group, rather than the individual. This was characteristic of most of the dancing scenes (Garfinkel 2003).

Tell Halula

A proto-historic tell site on the Euphrates River in Syria with various periods of occupation. In the Pre-Pottery Neolithic B village of the 9th millennium BCE, a number of painted plastered floors were found, with painted scenes of dancing females, one of which has been published (**fig. 2**; Molist 1998; Garfinkel 2003, 114–116). Dozens of Pre-Pottery Neolithic B sites are known in the Near East and in many of them houses with plastered floors were found.

Yosef Garfinkel: Archaeology of Dance

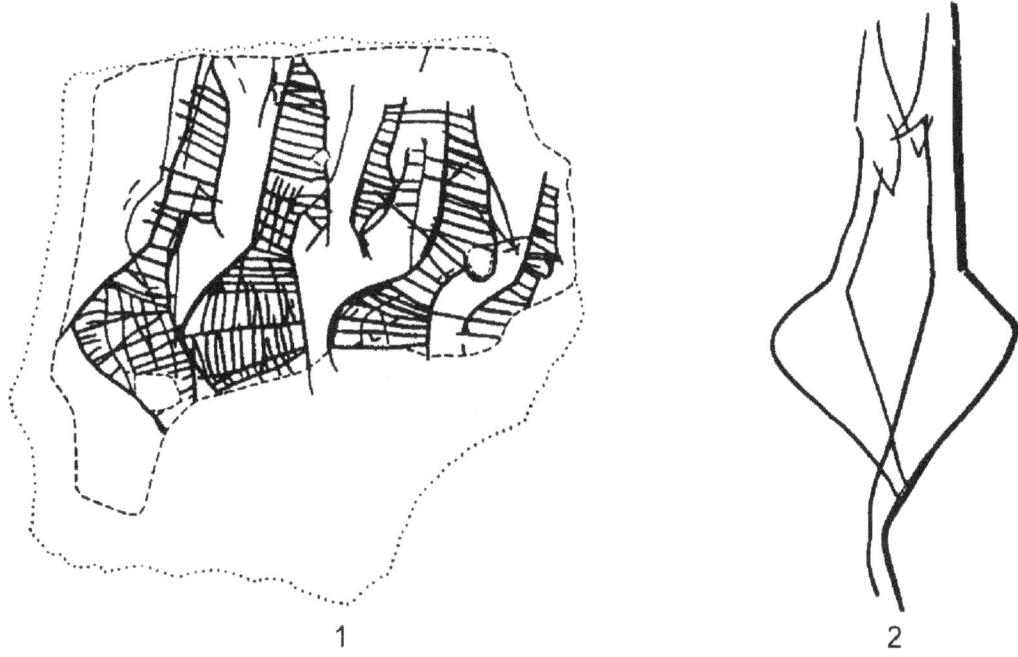

Figure 1. Prehistoric dancing scenes from the Upper Paleolithic site of Gönnerdorf in Germany (Bosinski and Fischer 1974, Pl. 59:87b, Fig. 36:78).

Figure 2. Prehistoric dancing scene from the Neolithic site of Tell Halula in Syria (Molist 1998, fig. 4).

However, floors painted with dancing figures are known only from Tell Halula. Why were these scenes painted at this site? Are there other aspects of the site's material culture that may explain this unusual phenomenon?

At both Gönnerdorf and Tell Halula, a large number of dancing scenes were discovered. Why were so many dancing figures depicted at these sites while they were completely absent at numerous contemporary sites? Probably these two sites were cultic centres in which specific ceremonies took place with the agglomeration of population from a large territory. One of the main characteristics of ritual is the exact repetition of various actions by the entire community: all assemble at the same location, at the same time, with the same dresses or masks, eating and drinking the same type of food, praying or repeating the same text, performing the same bodily gestures or dances. This is the strength of the ritual ceremony, to create unity across the entire community. The gathering of the community into the same place over the years for cultic ceremonies is reflected in the archaeological record by the accumulation of many dancing scenes in the same location.

At Gönnerdorf the representations include only young female figures, which may indicate the initiation rites of girls at the site. In traditional societies, initiation rites are performed separately for males and females. One is left to wonder where the people of the Gönnerdorf area conducted the initiation ceremonies for the boys.

At Tell Halula the general outline of the figures is much heavier, and they seem to represent mature women with prominent buttocks. If so, the published depiction may represent another type of female ritual, from a more advanced stage in life. It is also possible that the artistic convention of the Neolithic Near East presents female figures with prominent buttocks, and it was not considered a characteristic of age. If so, the Tell Halula scene may also represent initiation rites of girls. It should be noted that we do not have contemporary depictions of male ceremonies for the Magdalenian culture in Europe, or the Pre-Pottery Neolithic B of the Near East. Probably males performed other types of initiation rites, which were not documented.

Until now archaeologists were relatively passive concerning the discovery of dancing scenes, which were found accidentally during excavations. However, an archaeologist can become more active and initiate expeditions to sites or areas where dancing scenes were previously found, and thus uncover more. Multiplying the number of dancing scenes will aid a clearer understanding of the dance activities of that specific period or region. Many legitimate research questions about dance justify fieldwork at specific sites. For example, the sites of Khazineh, Tepe Musiyan and Tepe Sabz in southern Mesopotamia, on the Deh Luran Plain of western Iran, produced many dancing scenes (**fig. 3**; Gautier and Lampre 1905; Hole et al. 1969; Garfinkel 2000; 2003, 164–170), and it would be only natural that an archaeologist who is interested in studying dance would conduct new excavations at these sites.

Contextual study of the location in which a dancing scene was found

The context of an object that bears a dancing scene is clearly an archaeological element which can add much to our understanding of the dancing activity. Two different contexts will be discussed here: temples and graves.

Dancing scenes from temples

A large number of dance depictions were found in temples in the southern Levant from the Bronze and Iron Ages:

Megiddo

Three dancing scenes were engraved on stone slabs in the courtyard of the temple of Layer XIX, from the Early Bronze Age I, dated to the second half of the 4th millennium BCE (**fig. 4**; Loud 1948, Pls. 271:1, 272, 276:6; Garfinkel 2003, 282–284). Another engraving depicts a musician, a standing figure playing a lyre (Loud 1948, pl. 273:5). The combination of dance and music strongly suggests that these were part of the cultic ceremonies that took place in the temple, probably in the courtyard itself.

Tell Qasile

In a Philistine temple dated to the 12th–11th centuries BCE, a rounded clay cult stand depicting four human figures was found (**fig. 5:1**; Mazar 1980, 87–89, fig. 23). The figures are in a circle, in profile, holding hands. They were equidistant from each other, thereby creating a circle of dancing figures. Two musical instruments found in the temple complex were conch shells, known as triton or trumpet shells (Mazar 1980, 118). A broken conch was found in Layer XII, at Locus 275. This area is part of a large open area in front of the temple (Mazar 1980, fig. 48). Dancing activities probably took place in this courtyard, and a broken musical instrument was left there as it could not be used anymore. A complete item, which still works today (Mazar 1980, 118), was found in Layer XI, at Locus 227, a storage room inside the temple. Since it still worked, the item was not discarded, but was kept in the storage room for further use. In Layer X the temple was enlarged at the expense of the courtyard which became much smaller, and was no longer functional as a dancing ground. No musical instruments were found in Layer X and it is possible that both music and dance disappeared altogether from the temple cult.

Kuntillet 'Ajrud

An Israelite cultic centre in Sinai, dated to the early 8th century BCE, Kuntillet 'Ajrud had both cultic inscriptions and depictions of various ritual scenes (Meshel 1993; Beck 1982). A large pottery storage jar was painted with a scene of five human figures standing in a row facing left (**fig.**

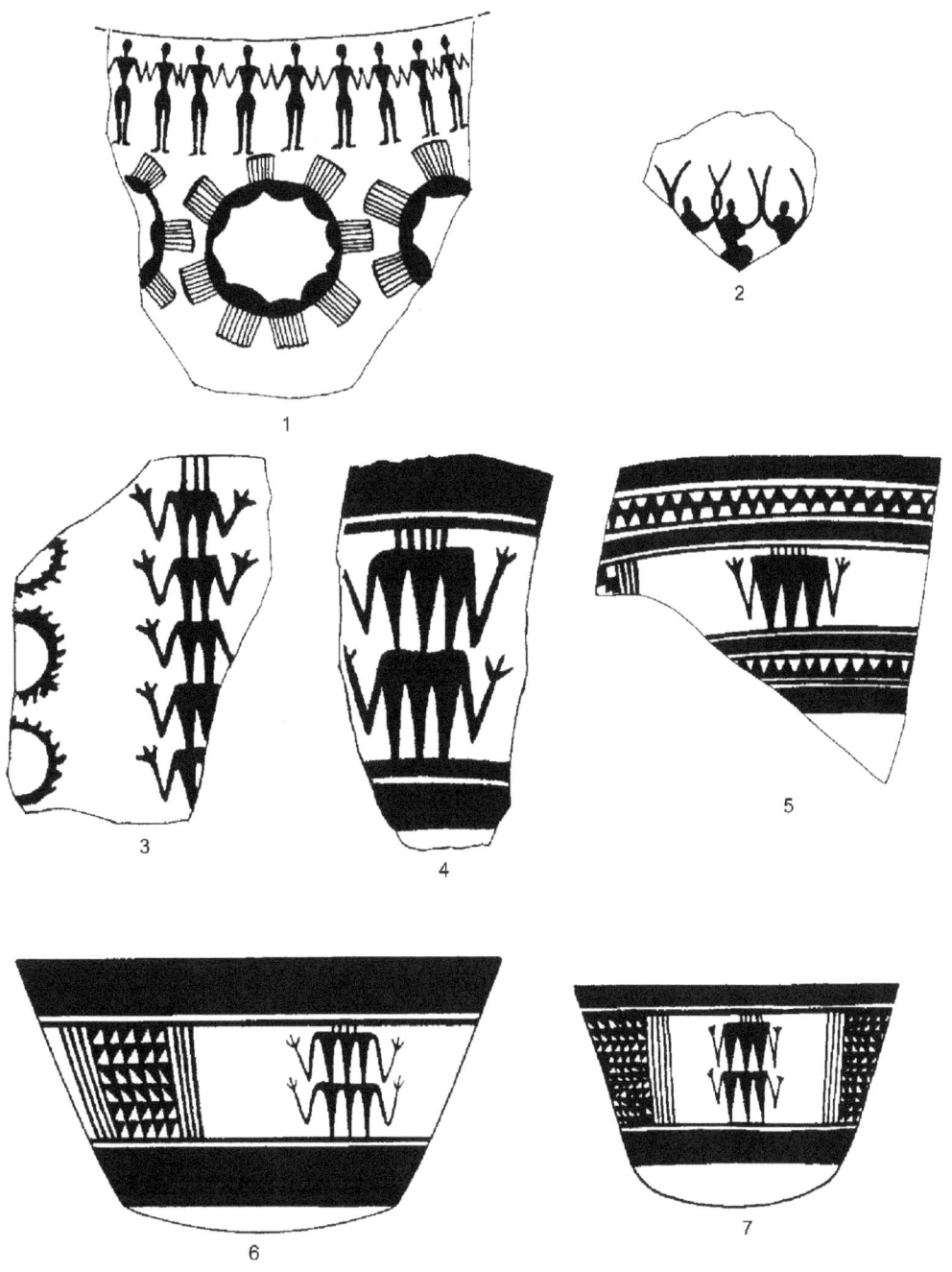

Figure 3. Prehistoric dancing scenes from Early Chalcolithic sites in western Iran: Khazineh, Tepe Musiyan, Tepe Sabz (Gautier and Lampre 1905, Hole et al. 1969, Garfinkel 2000, Garfinkel 2003, 164–170).

5:2; Beck 1982, fig. 3). Their arms are bent as if they were praying, and thus were designated in the publication as worshippers. It is possible, however, that they are clapping their hands. They are not shown as a dynamic group of dancers, so they may represent a procession. But as the entire artistic assemblage of the site consists of rather crude graffiti rather than the product of trained artists, it is possible that this scene also depicts dancing. Another drawing at the site depicts a musician, a seated woman playing a lyre (Beck 1982, fig. 5). The combination of procession/dance and music clearly indicates that these were part of the cultic ceremonies that took place at the site.

The site of Kuntillet 'Ajrud is a rectangular enclosure of c. 15 by 25m, with a large open courtyard at its centre, and this courtyard probably functioned as the dance ground for the site.

Qitmit

An Edomite open cult centre on top of a hill in the Beersheba valley, it is dated to the late 7th century BCE (Beit-Arieh 1995). The site includes several structures and a large circular open enclosure (**fig. 6**, Locus 114). Among the rich assemblage of hundreds of cultic paraphernalia is

Figure 4. Dancing scenes from an Early Bronze Age temple in Megiddo (Loud 1948, pls. 271:1, 272, 276:6; Garfinkel 2003, 282–284).

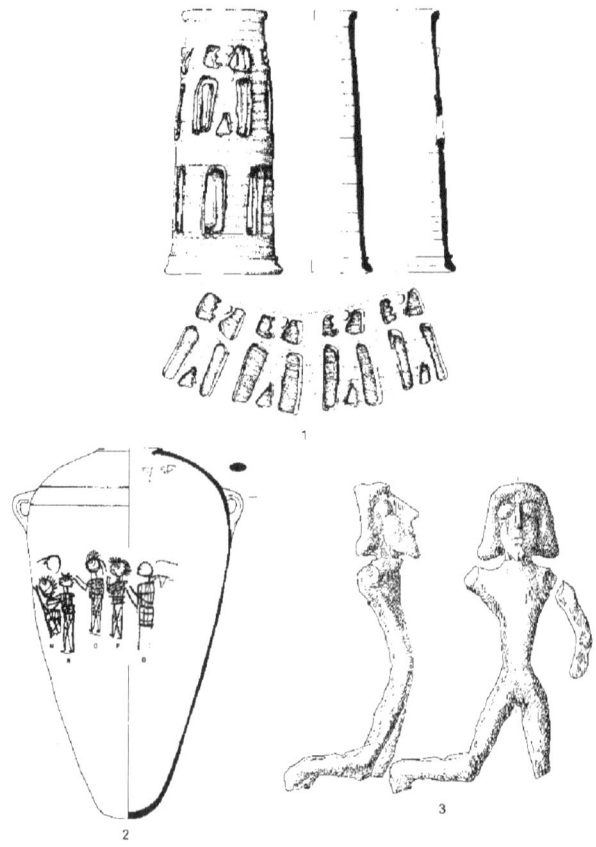

Figure 5. Dancing scenes from Iron Age Tell Qasile (Mazar 1980, fig. 23), Kuntillet 'Ajrud (Beck 1982, fig. 3) and Qitmit (Beck 1995, fig. 3.79).

a clay male figurine shown in a dancing posture (**fig. 5:3**; Beck 1995, 110–111). In addition, clay figurines with three different musical instruments were found: six tambourines, a lyre and a double-pipe (Beck 1995, 161–168). Here too a combination of dance and music were clearly part of the cultic ceremonies.

The open circular enclosure at Qitmit (Locus 114) contained few finds, and the excavators had reservations regarding its function. Originally they called it a pen for livestock. However, after uncovering a bench attached to the southern side and a large stone (massebah?), they understood it to have a cultic function as well (Beit-Arieh 1995, 24). However, the rich cultic paraphernalia of the site was found elsewhere, and this area may have been used for circle dancing. The bench could have been for the musicians, who played the tambourine, lyre and double-pipe.

The connection between dance and ritual has been discussed often (see, for example, Sachs 1952, Lange 1976; Hanna 1987). It is not surprising therefore that dancing is depicted at various temples in ancient Israel, from Early Canaanite Megiddo to Philistine, Israelite and Edomite sites. In each there is evidence for music. Sometimes the musical instruments themselves were found (such as at Tell Qasile), but usually there were represented by depictions of people playing them. These representations were made using various techniques: engraving (Megiddo), painting (Kuntillet 'Ajrud) or figurines (Qitmit). All four sites had large open areas where dancing could be performed. At Qitmit this area was marked by a round wall, so the dancing ceremonies each time were performed in exactly the same confined location.

Circling is an important component in religious rituals and has magical connotations. Biblical Jericho was circled seven times before it was conquered and in Jewish wedding ceremonies the bride walks round the bridegroom seven times. Circle dancing was an important component of the ritual activity performed in these temples by both the local people and also the distant population who made pilgrimages to the holy place. Indeed, even today, pilgrimages to holy places in different religions includes the encircling of a holy object, building or grave: Moslems encircle the Ka'aba in Mecca seven times, Buddhists encircle the stupa usually three times (Akira 1987, 95), Hindus encircle temples or inner courts (Eck 1987), and rare cases occur even in Christianity, such as the case of Saint Patrick's Purgatory in Ireland where the basilica is encircled four times (Eck 1987).

As discussed above, a major characteristic of ritual is the exact repetition of various acts by the entire community. If a community determined that dances would always be

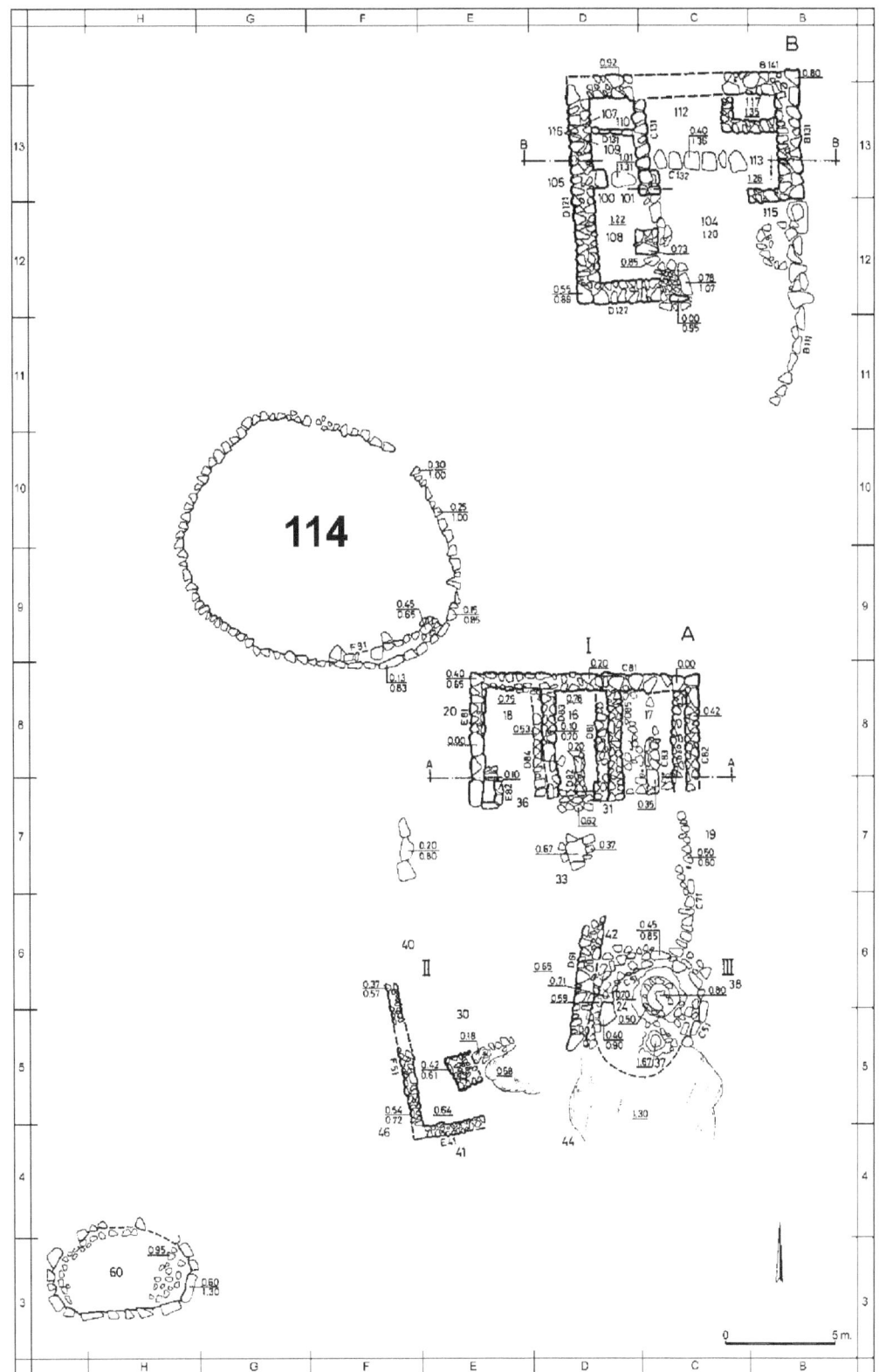

Figure 6. The hill-top cultic complex of Qitmit (Beit-Arieh 1995, fig. 1.6). The rounded open enclosure (Locus 114) may represent a dancing ground for circle dancing.

performed in the exact same location, there would be a need to mark the place. If a confined temple courtyard, like those at Megiddo and Tell Qasile, was not focused enough, another rounded area could have been designated. Such an example was found at the rounded enclosure of Qitmit, and is also known from rounded stone platforms found at Knossos on Crete (Warren 1984). Marking the exact dance location is indeed an unusual phenomenon, perhaps associated with a local mythological tradition that a god or holy saint once danced at that specific spot.

Dancing at burials

Many dancing scenes have been discovered at grave sites. The connection between dance and burial ceremonies is well attested in ethnography, linguistics and in dancing scenes at burial ceremonies. Mourning dances have been reported in various ethnographic researches (see, for example, Sachs 1952, 74–75; Morphy 1994). From a linguistic point of view, the Semitic root *rqd* is used in Hebrew to describe dancing, while in Syriac it is used to describe mourning (J. Greenfield, pers. comm.). The common semantic field demonstrates that in the ancient Semitic milieu funeral services included dancing. Mourning customs are described in Hebrew by the root *spd*, and it is interesting to note that the two roots *rqd* and *spd* appear together as opposites in Ecclesiastes 3:4 - "A time to mourn (*spd*) and a time to dance (*rqd*)".

Contextual analysis associates dancing activities with burials. In ancient Egypt mourning dances appeared on tomb wall paintings (Brunner-Traut 1958, 61). In the Mycenaean and Philistine cultures of the second half of the second millennium BC mourning female figurines were applied to the rims of round clay vessels (Iakovidis 1966; Dothan 1982, 237–245), indicating that the ceremony was carried out in the form of a circle dance. These items were found in graves (Iakovidis 1966). Elaborate depictions of burial ceremonies were often painted on large burial vessels of Greek Geometric pottery. Here we see a clear connection between the function of the vessel and its decoration. Sections of the scene include friezes with rows of male or female figures in formalized body gesture which indicate rhythmical movement in procession or dance (see, for example, Schweitzer 1971, 37–52, pls. 30–36, 40–41, 46–50).

Two items demonstrate that mourning dances probably existed in the proto-historic Near East.

'Ein el-Jarba

This site is located near Haifa, Israel, and dated to the 6th millennium BC. A complete holemouth jar was found near a burial (**fig. 7:1**; Kaplan 1969; Garfinkel 2003, 155–157). On it two large human figures were painted on both sides. The (masked?) heads are in profile, facing left, thus indicating clockwise movement around the jar.

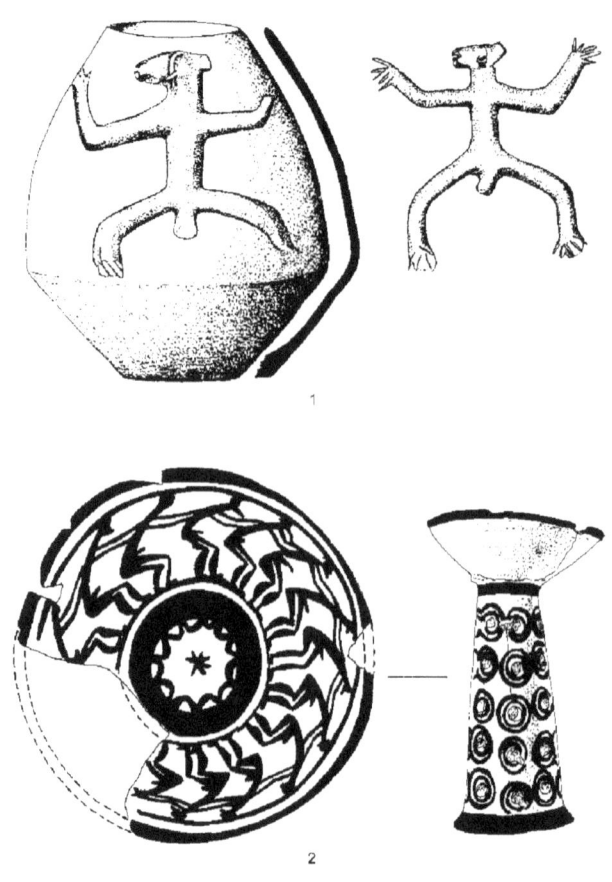

Figure 7. Dancing scenes from prehistoric burials in the Near East: 1. 'Ein el-Jarba (Kaplan 1969; Garfinkel 2003, 155–157), 2. Tall-i Jari A (Vanden Berghe 1952; Garfinkel 2003, 201).

Tell i Jari A

This site is located near Persepolis in Fars, Iran, and dated to the first half of the 5th millennium BC. A complete bowl on a tall fenestrated pedestal was found in a grave (**fig. 7:2**; VandenBerghe 1952; Garfinkel 2003, 201). The bowl is decorated by a circle of thirteen identical naked male figures in profile, each one touching the shoulders of the figure in front of him with both hands, moving clockwise.

Dancing figures, when depicted in profile on rounded objects, moved clockwise or counter clockwise. In the proto-historic Near East, people danced in a counter clockwise direction (Garfinkel 2003, 44–47). However, at 'Ein el-Jarba and Tell i Jari A, the figures move clockwise, opposite to the conventional direction. It is possible that in funerary ceremonies, people moved in the direction opposite to that used in other ceremonies. A similar phenomenon was reported during pilgrimages to the second temple in Jerusalem (Danby 1938: Middoth 2, 2). While the community circled the temple from right to left, a counter clockwise movement, mourning families went from left to right, a clockwise movement around the temple.

Dance accessories

The most eye-catching feature of many festivals and dances of traditional societies are the lavish and colourful dresses, masks, and other dance accessories. Ethnographic examination has revealed that sometimes enormous efforts were devoted to the manufacture of these items. If not made of perishable materials, these objects are often uncovered by archaeologists. However, associating an item with dance activities is not self-evident – the context of the find must be established.

A few years ago, a zooarchaeological study at Çatal Hüyük in Turkey suggested that worked bones from the spread wing of a crane were used in dances. This part of the wing has very little flesh, but large flight feathers. The cut marks on the bones indicate that this it was not a result of butchery, but probably prepared. This wing could have been attached to the shoulder of a dancer and used as a costume impersonating cranes (Russell and McGowan 2003). Depictions of cranes were also found at the site, showing that this bird had a meaningful symbolic significance for this community.

Usually dances are performed in groups, so one would expect that several dancers would perform a crane dance together. This was found to be the case by the reconstruction accompanying the article (Russell and McGowan 2003, fig. 6). Thus, if a large number of crane wing bones were to be found together, it would strengthen this interpretation, but if each dancer organized his own costume, the wing bones need not be found together. The Çatal Hüyük example demonstrates that archaeological evidence for dance can be retrieved from unexpected sources.

Another dance accessory is the mask. Many studies have been devoted to masks, but these usually concentrate on their typology and dating. We need a study on masked performances, which will emphasize the role of the mask in dances. A few years ago I started to work on this aspect but this study had not yet been completed.

Summary

Archaeology of Dance is part of a larger field which can be designated the "Archaeology of Ritual Performance". Rituals and ceremonies are elaborate events with a complex set of actions, involving talking (praying, blessing, storytelling), eating (drinking, feasting), bodily gestures (clapping hands, putting one hands on other people heads) and movement (dancing, moving in procession, circling). Dance is not done in isolation, but as part of a more complex ritualistic activity. While feasting does leave clear and direct archaeological remains (see, for example, Dietler and Hayden 2001; Goring-Morris and Horwitz 2007; Ben-Shlomo et al. 2009), dancing leaves only elusive archaeological evidence.

One direction for the Archaeology of Dance to take is to uncover more dance depictions and to study their context. Dance in temples can add another dimension to the archaeology of cult, which tends to focus on frozen aspects like architecture and cultic paraphernalia, but neglects to look at the human activities such as music and dance. It is quite possible that large temple courtyards were used for dance. Another promising source is burial customs where the depiction of dance can add information on burial rites.

A second path open to the Archaeology of Dance is to identify and analyse dance accessories. Rich and extravagant dresses and masks can be uncovered during excavations, as demonstrated by the case study from Çatal Hüyük. Interdisciplinary research combining zooarchaeology and art history enabled the discovery of a possible crane dance. Only combined efforts will enable breakthroughs in studying the dance in past societies.

Indeed, the study of the Archaeology of Dance is difficult but not impossible. In this short presentation I have examined how archaeologists can identify dance activity in the archaeological record. While the evidence is not always obvious, the subject matter should not be overlooked altogether. With more awareness, relevant data will be recognized and collected to create a better understanding of dance activities in the past.

References

Akira, H. 1987. 'Stupa Worship', in M. Eliade (ed.), *Encyclopedia of Religion*, Vol. 14. New York, MacMillan, 92–96.

Beck. P. 1982. The Drawings from Horvat Teiman (Kuntillet 'Ajrud), *Tel Aviv* 9, 3–68.

Beck. P. 1995.'Catalogue of Cult Objects and Study of the Iconography. In I. Beit-Arieh (ed.), *Horvat Qitmit. An Edomite Shrine in the Biblical Negev,* 27-197. Tel Aviv university, Institute of Archaeology.

Beit-Arieh, I. 1995. *Horvat Qitmit. An Edomite Shrine in the Biblical Negev*. Tel Aviv university, Institute of Archaeology.

Ben-Shlomo, D., Hill, A.C. and Garfinkel, Y. 2009. Feasting Between the Revolutions: Evidence from Chalcolithic Tel Tsaf. *Journal of Mediterranean Archaeology* 22, 129–150.

Biesele, M. 1978. Religion and Folklore. In P.V. Tobias (ed.), *The Bushmen. San Hunters and Herders of Southern Africa*, 162-172. Cape Town and Pretoria, Human and Rousseau.

Bosinski, G. 1970. Magdalenian Anthropomorphic Figures at Gönnerdorf (Western Germany). Preliminary Report on the 1968 Excavations. *Bolletino del Centro Camuno di Studi Preistorici* 5, 57–97.

Bosinski, G. *1979. Die Ausgrabungen in Günnersdorf 1968–1976 und die Siedlungshefunde der Grabung 1968. Wiesbaden*, Franz Steiner Verlag.

Bosinski, G. and Fischer, G. 1974. *Die Menschendarstellungen von Göennersdorf der Ausgrabung von 1968*. Wiesbaden, F. Steiner.

Brunner-Traut, E. 1958. *Der Tanz im Alten Ägypten, Nach bildlichen und inschriftlichen Zeugnissen*. Ägyptologische Forschungen Heft 6. Gluckstadt, Verlag J.J. Augustin.

Danby, H. 1938. *The Mishnah*. Oxford, University Press.

Dietler, M. and Hayden, B. 2001. *Feasts: Archaeological and Ethnographic Perspectives on Food, Politics and Power*. Washington, Smithsonian Institute Press.

Dothan, T. 1982. *The Philistines and Their Material Culture*. Jerusalem, Israel Exploration Society.

Eck, D. L. 1987. Circumambulation. In M. Eliade (ed.), *Encyclopedia of Religion*, Vol. 3, 509-514. New York, MacMillan.

Garfinkel, Y. 2000. The Khazineh Painted Style of Western Iran. *Iran* 38, 57–70.

Garfinkel, Y. 2003. *Dance at the Dawn of Agriculture*. Austin, Texas University Press.

Garfinkel, Y. 2010. Dance in Prehistoric Europe. *Documenta Praehistorica* 37, 205–214.

Gautier, J. E. and Lampre, G. 1905. Fouilles de Moussian. *Mémoires de la Délégation archéologique en Perse* 8, 59–148.

Goring-Morris, N. and Horwitz, L. K. 2007. Funerals and feasts during the Pre-Pottery Neolithic B of the Near East. *Antiquity* 81, 901-919.

Gruber, N. I. 1981. Ten Dance-Derived Expressions in the Hebrew Bible. *Biblica* 62, 328–346.

Hanna, J.L. 1987. 'Dance and Religion'. In M. Eliade (ed.), *Encyclopedia of Religion*, Vol. 4, 203-212. New York, MacMillan.

Hole, F., Flannery, K. V. and Neely, J. A. 1969. *Prehistory and Human Ecology of the Deh Luran Plain. An Early Village Sequence from Khuzistan, Iran*. Memoirs of the Museum of Anthropology, No. 1. Ann Arbor, University of Michigan.

Iakovidis, S. P. E. 1966. A Mycenaean Mourning Custom. *American Journal of Archaeology* 70, 43–50.

Kaplan, J. 1969. Ein El-Jarba. Chalcolithic Remains in the Plain of Esdraelon. *Bulletin of the American Schools of Oriental Research* 194, 2–39.

Katz, R. 1982. *Boiling Energy. Community Healing Among the Kalahari Kung*. Cambridge, Harvard University Press.

Lange, R. 1976. *The Nature of Dance. An Anthropological Perspective*. New York, International Publications Service.

Marshall, L. 1969. The Medicine Dance of the !Kung Bushmen. *Africa* 39, 347–381.

Mazar, A. 1980. *Excavations at Tell Qasile. Part 1: The Philistine Sanctuary: Architecture and Cult Objects*. Qedem Monographs 12. Jerusalem, Institute of Archaeology, the Hebrew University of Jerusalem.

McNeill, W. H. 1995. *Keeping Together in Time. Dance and Drill in Human History*. Cambridge, Harvard University Press.

Molist, M. M. 1998. Des représentations humaines peintes au IXe millénaire BP sur le site de Tell Halula (Vallée de L'Euphrate, Syrie). *Paléorient* 24/1, 81–87.

Morphy, H. 1994. The Interpretation of Ritual: Reflections from Film on Anthropological Practice. *Man* 29, 117–146.

Royce, A.P. 1977. *The Anthropology of Dance*. Bloomington, Indiana University Press.

Russell, N. and McGowan K. J. 2003. Dance of the Cranes: Crane Symbolism at Çatalhöyük and Beyond. *Antiquity* 77, 445–455.

Sachs, C. 1952. *World History of the Dance*. New York, Seven Arts.

Schweitzer, B. 1971. *Greek Geometric Art*. London, Phaidon.

Vanden Berghe, L. 1952. Archaeologische opzoekingen in de Marv Dasht vlakte. *Jaarbericht Ex Orient Lux* 12, 211–220.

Warren, P. 1984. Circular Platforms at Minoan Knossos. *Annual of the British School at Athens* 79, 307–323.

WHO IS THE ANCIENT EGYPTIAN DANCER?

Batyah Shachter

The Jerusalem Academy of Music and Dance,
The Hebrew University of Jerusalem

Abstract: Any attempt to define the notion of 'dance' or 'dancer' is fraught with difficulties; this is the case even today, when we come to examine dance that we can witness or execute. This renders the study of dance in ancient Egypt all the more problematic, since the only documentation we have about dance in ancient Egypt is only that of plastic representations; as such, defining notions of 'dance' or 'dancer' is largely a matter of speculation and interpretation. This is further complicated by the fact that ancient Egyptian documentation of dance and dancers was not intended to record realistic information. Rather, it had to submit to clear rules of representation, presenting sufficient information for viewers to identify and recognise events known to them. This paper will attempt to formulate a definition of dance, while describing the principles of ancient Egyptian representation of the body and movement; using these tools, I will try to provide an analysis of dance scenes in ancient Egyptian art.

Foreword

I have written this paper from the perspective of a dancer who has been involved in dance, the study of movement, and the creation and performance of dances for many years. When considering dance in ancient Egypt, I examine two-dimensional, silent images which provide no real clues regarding the movement which inspired the picture. Moreover, as a dancer trained to observe movement itself, many of the figures in Egyptian pictures and reliefs appear, to me, to be dancing. For example, a man bending down to fish, or kneeling in order to slaughter an animal (**fig. 1**), is represented in a deliberate and studied movement. If we were to separate the clearly identifiable act performed by the man from the pose in which we see him, we could see a dancer in the middle of a rehearsed, trained and deliberate movement. At the same time, there are images where there is no doubt that the figures are performing a dance which has no additional purpose (**fig. 2**). Originally, I identified all the figures who were not obviously performing any specific action as a dancer. However, after consideration, and the need to find out more about the place, nature, function and character of dance in ancient Egypt, I realized that it was necessary to define what was and was not dance within the images of human figures so common in Egyptian art.

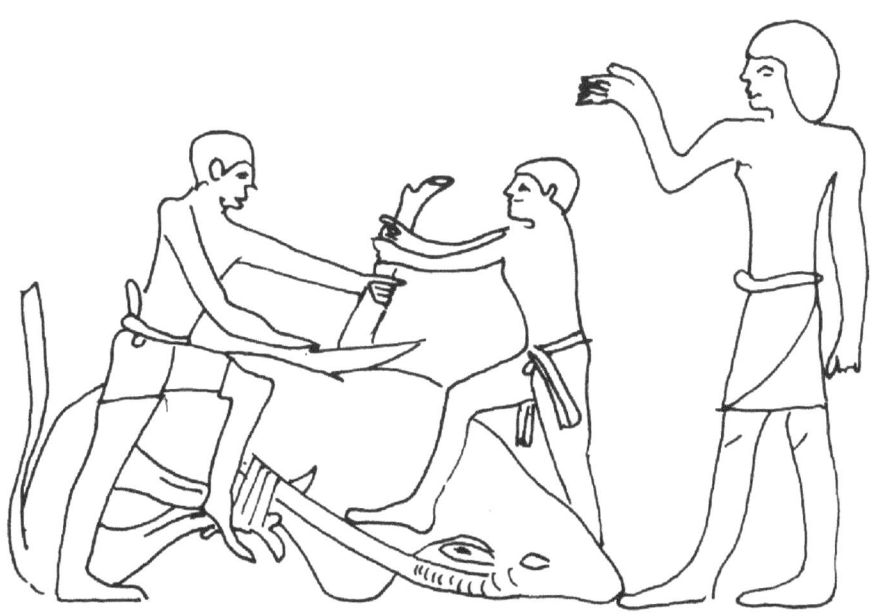

Figure 1. Priest standing on the right next to a scene of slaughtering, tomb of Ninetjer, Giza, Old Kingdom (after Junker 1951, fig. 46).

Figure 2. New Kingdom, tomb of Kheruef, 192, Thebes (after Strouhal 1992, 47).

In this paper I intend to try to define which amongst the figures drawn in movement in Egyptian art can be identified as dancers, which cannot, and which are doubtful. Each definition will be based upon observations of the movements portrayed in the imagery. It will be analyzed, using the rules of the Egyptian canon concerning the representation of the human body and my own understanding of the movement of the human body through observation and execution.

This approach has not previously been considered in Egyptological research. The few scholars who have engaged with the subject concluded that certain figures were dancers of one type or another, and dealt with them in this context. Emma Brunner-Traut in her book *Dance in Ancient Egypt* (1958) distinguished between dancers and those engaged in sports activities and games, and she also differentiated dance from the organization of groups in religious processions. She saw dance as "rhythmical movement that springs from the need for artistic or religious expression or a spontaneous and obvious expression of life" (Brunner-Traut 1958, 9). However, why these figures are identified as dancers and others as worshippers, gymnasts or maid-servants serving food at a feast, what characterizes these differences, according to which criteria is the question itself investigated – these questions have not so far been asked in the literature concerning dance in ancient Egypt.

In order to investigate these problems I shall give a detailed presentation of the issues regarding the definition of dance, as discussed by contemporary dance practitioners. I will consider the fact that dance is an action which exists in time, and the difficulties which derive from this for the documentation, preservation and even repetition of "the same dance" (McFee 1995, 93). Later, I will consider the principles of the Egyptian canon concerning the portrayal of the human figure in painting and the way Egyptian artists dealt with the problem of a two-dimensional representation of two- and three-dimensional space. These two elements are of the utmost importance when attempting to understand or examine movement in paintings. One issue is the manner in which the human body is represented in Egyptian art in general. Although the human figure is drawn naturalistically and can be identified without any doubt, the manner in which it is represented is not realistic with regard to the accuracy of the shape of the body and its movement in space (Lexova 1935, 16). Finally, I will examine the features that characterize dance in Egyptian art by considering examples according to the elements presented beforehand, including the limitations which result from the minimal knowledge available to us. To me, this process is the first step in the study of Egyptian dance and the contemplation of the different possibilities of movement which exist within it. This will also serve as a base for further study of the social, artistic, textual and religious contexts of ancient Egyptian culture.

How can dance be defined?

What turns one sequence of movements into dance when another sequence is not identified as such? What differentiates the sportswoman in the high-jump from a group of people moving within the defined space of the stage, performing a sequence of actions dictated in advance, with or without musical accompaniment and with or without an audience (McFee 1995, 49-54)? Questions of the definition of dance have preoccupied the creators of dance themselves since the middle of the twentieth century, when dance performance moved from the theatrical stage and into the street, the studio and other various spaces. During this period the question of the trained and professional dancer also fell under the scrutiny of a group of artists who preferred to utilise people who had not undergone systematic training as dancers and did not have any traditional or conventional technique (Scott 1997, 7). If the ability to define a dancer or a dance performance is complex and multi-faceted when referring to dance or movement that can be watched in the here and now, this issue becomes even more acute when applied to ancient images from ancient times, whose rules and customs are known to us only from the pictures and texts that have survived (Garfinkel 2000, 16).

Determining the definition of dance is a complex task, as what we want to define is composed of elements of movement (McFee 1995, 52, 55-6). In practice, we have to distinguish between the different sequences of movement and sort them into categories in order to decide which are dance and which are not - to make this distinction we are forced to analyze every area of our lives from the perspective of movement. If we adopt the broadest definition, we can, in fact, view every movement of the human body as dance. But since I wish to deal with the subject of dance as it is recognized and expressed in human society, I must find a definition that identifies and distinguishes it from other sequences of movements (McFee 1995, 49). Therefore, I will not consider a wide-ranging definition in which every movement performed could be considered dance.

The concept of dance is inherently understood by everyone and does not require an explanation or definition to identify this action, whether it is dance on the stage of a theatre, folk dancing in the street, ballroom dancing, or dance in religious ceremonies. In all of these cases the person watching will recognize that they are viewing a dance (Alter 1996, 4).

Understanding these two extremes - the known and identified on the one hand and the complexity of definition on the other - I will try to illustrate the relevant points which can help identify dance as a concept, and which reveal something of the debate around this question in the dance world. I do not presume to give a complete picture of the current debate around the definition of dance, but to raise the points that, to the best of my understanding, concern the subject of this paper.

Dance is movement for the sake of movement itself

One basic distinction defines dance as movement that is performed for its own sake and not in order to perform another action (McFee 1995, 51). This distinction excludes all the actions of daily life in which movement is executed in order to achieve some goal external to the movement itself. That includes both simple everyday actions such as speaking, eating and walking, and complex movements that require concentration and clear intention such as playing musical instruments, sculpting, building, etc. It may be said that, by their very nature, these activities are not dance (McFee 1995, 55-6). The movements that serve them may, in themselves, be extremely complex or performed automatically, but ultimately they are the means for the fulfilment of another goal. In these cases, the importance of movement is that it is performed in order to bring about a desired, practical result. Therefore, dance may be defined as an activity in which the execution of movement is the centre of attention rather than any other functional outcome that occurs as a result of it. It can be said that, in dance, movement functions in order to create itself and, as such, exists as a sequence of movements of the human body in a certain space and at a given time (Alter 1996, 7-8).

From this we can conclude that dance is an activity in which movement is itself a tool and a goal at the same time. In this case the question must be asked: why did an activity whose goal is its implementation develop in human society? A record of the existence and importance of dance in human society has existed from prehistoric times (Garfinkel 2000). Dance managed to fulfil and express the different needs of man in his social life. In human culture, dance has been used in many ways and for the greatest variety of purposes - as a social expression of joy or grief (wedding dances and mourning dances at funerals), religious ceremonies, prayer (dances of supplication for rain), play, pleasure (ballroom dancing or folk dancing), for some type of social bonding (war dances), and for artistic expression. More recently, trends have developed in which dance serves as a means of self-expression, release and the understanding of the human psyche and its healing. From this we can deduce that the movement of the human body has served and still serves as a tool to satisfy various personal and social needs that can be understood and defined in different ways (Alter 1996, 17, 19).

What is and what is not dance?

When dance serves so many purposes and has so many different aspects, how can we distinguish between them or decide if a certain sequence of movements is a dance or not? Whilst we define dance as an activity in which movement is the goal and the means at the same time, we are not denying the possibility that any movement can be used in the performance of the dance itself. "It is at least possible... that every movement that occurs in dance, may also exist in another context: there is no pattern of movement that can in itself be defined as dance" (McFee 1995, 51). Thus, for example, walking in itself is not dance, although dance can definitely include walking as one of its movements, when the purpose of walking is not moving from one place to another, but as part of the dance itself. In this case the emphasis will be on the manner in which the walking is performed, which would not be the case if it were just a function of "reaching" another place (McFee 1995, 49, 51).

In contemporary dance this question has come to the fore, mainly because many of the conventions concerning movement that we are accustomed to seeing as dance have been overturned by choreographers who prefer to create dances that incorporate everyday movements, performed by people not trained as dancers according to accepted tradition (McFee 1995, 67). This is relevant since it sheds light on the question as to what dance is in its widest sense, and does not rely solely on the simple and accepted identification of the dancing figure (McFee 1995, 49-54). Once again, the question arises: what defines an activity as dance? McFee[1] claims that in fact there is no need or possibility to provide a definition but instead we should try to find the differences between dance and other actions similar to it, such as gymnastics (McFee 1995, 49). He also claims that the only thing that defines this difference is the way it is viewed by the audience and the criteria by which it is judged (McFee 1995, 50, 66). Thus, for example, in a gymnastics competition the audience will consider the height of the jumps, the accuracy of the landing, or the speed at which a routine progresses. In contrast, a dance performance is judged according to artistic criteria, the expression of the dancer, the accord between the dance and the music (if there is any), the harmony or dissonance of the entire dance, etc. They are viewed differently, with different expectations, and these expectations are what determine that one is dance and the other gymnastics on the part of both the performer and the spectator (McFee 1995, 58-9, 65-66). McFee also notes that external conditions such as costumes, lighting and a stage do not necessarily define the

---

[1] In his 1995 book "Understanding Dance", McFee deals with the definition of dance and philosophical questions which are connected with and follow from this discussion.

activity as dance since they can represent other activities that take place with their assistance (such as a play).

Dance as a performing art which exists in time

Dance is an art form which exists in time and as such it exists for one moment and disappears in the next. All our efforts to capture, document and preserve dance are limited to its description rather than its manifestation. This debate about dance still exists today when it can be filmed by different methods, and yet photography is not performance and the screen is not the stage (Alter 1996, 14). The problem has implications for dance which is documented only in drawings, and as such provides us with very little information for the investigation of ancient dance using the minimal knowledge in our possession.

In practice, it is impossible to "catch hold" of the dance in order to watch it again. Even if we were in the same room and the same dancer repeated the same choreography, we would never see the same dance (McFee 1995, 93-94). The dancer might be more tired, more excited, more relaxed, etc. We can therefore see that it is impossible to view "the same dance" over and over again. This would be even clearer if another dancer were to dance the same steps at the same rhythm. With a different physique, a different inner rhythm and a different interpretation, the dance would change from one person to another and from one moment to another. Even if we as spectators could identify what was similar, it still would not be a repeat of the event itself but rather an additional performance of the same movements. Consequently, it would be a different dance each time it was performed (McFee 1995, 93, 101).

Nowadays, we can film a dance as it is being performed which creates the deceptive impression that it is possible to watch dance on the screen. While it is true that the event is documented, it has many limitations. The camera only gives us a fraction of the entire space and the movement within it. Frequently, it reflects the desires and understanding of the cameraman concerning the space, movement and relative importance of the details. This experience is in no way similar to watching a live performance of dance, and usually only gives us a slight intimation of the dance itself and the experience it generates (Alter 1996, 14; McFee 1995:88). As spectators, the experience of viewing an existing work of art, such as a painting or a sculpture, changes repeatedly. We see things differently each time, if only because our personal and subjective feelings change from moment to moment. As to the performing arts (acting, dance, music), there are undoubtedly changes in the performer and the spectator (McFee 1995, 89, 101).

When we come to define dance we are dealing with an activity that cannot be documented or duplicated. Moreover, it is an activity that is difficult to define, being composed of elements that serve us in our daily lives - movements of the human body. One of the significant criteria for identifying dance is met when the movement exists for its own sake and not in order to execute some other functional goal. Even when we exclude all the instances that fall outside this definition, we still need to define what distinguishes the use of body in dance from that in sport, for example. Since the definition of dance can be very problematic if based solely on watching the movements, it is our point of view both as dancers and as spectators which create the criteria that will determine the activity or the event as dance. Our observation of the movements and the event taking place determines what the act is. If the dancer says that he is dancing and the audience understands that it is watching dance, then the way the dance is executed and the expectations of the spectator will together create the existence of dance (Scott 1997, 22 note 37). This dance event may include movements that the spectator identifies as typical of dance, recognizable and identifiable by its steps and movements; alternatively the spectator may associate the movements with those of everyday life or any other specific activity, or any combination of both.

Documentation of dance of ancient Egypt

From the extant Egyptian finds, no groups of three-dimensional figures have ever been found. While statues of Bes, the dancing god, considered to be a super-human figure (Spencer 2003, 112), have been found, as well as two statues of a long-haired female dancer in a back-bend position (Brunner-Traut 1958, 39), these three-dimensional representations appear as individuals and do not give us a picture of a group moving in space. Extant representations of group dances are two-dimensional, making it even more difficult to understand how they could have moved, and they provide few clues as to the spatial perception and group formation that was common in Egyptian dance. Since dance always takes place in space, in order to understand the possible nature of their movements, we have to learn some essential details of the way the human body was portrayed and the method of dealing with space according to the Egyptian principles of two-dimensional art.

Principles of representing the body and space in two-dimensional Egyptian art

Egyptian art is a semiological system in which a picture is a symbol of the message behind it (Tefnin 1984; 1991). Egyptian art does not accurately portray reality but rather represents, in codes, the scenes which were familiar to the viewers; the body and its movement is not portrayed with realistic accuracy but rather in an indirect manner that is simple, clear and effective for the artist's needs (Aldred 1986, 18). This is done using the aspective method, whereby the parts of the body are drawn separately in the manner which constitutes the most characteristic aspect of each element. The picture of the complete body is thus a composition of these parts (Shäfer 1986, 18).

This system creates codes and agreements concerning the content and the way it is portrayed. Egyptian art is in fact a language that connects the regime and the individual, and

Figure 3. Soldiers running, New Kingdom, El-Amarna, tomb of Meru (after Shäfer 1986, 179).

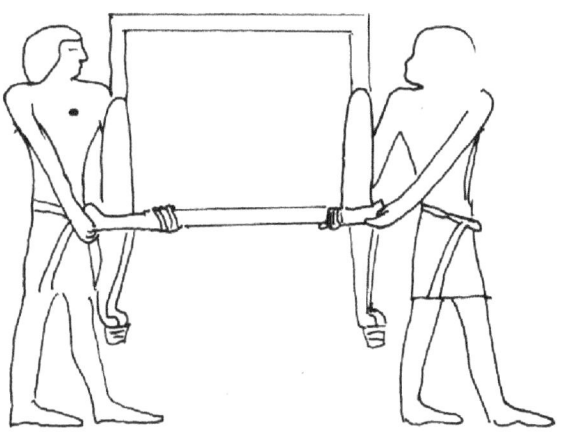

Figure 4. Two figures carrying a chair, tomb of Ti, Old (after Shäfer 1986, 143).

man and god. In the Egyptian language, there are several words and terms to represent the concept of dance (Green 1883; Brunner –Traut 1958; Montet 1925). When we look for the existence of dancers and dancing we are searching for the codes and symbols that characterize dance and portray it clearly. We have to remember that the events represented were clear to the Egyptians and the hints that were given provided the information necessary for them to recognize what the picture portrayed.

Unlike modern art, Egyptian painting was not a means of free expression. The Egyptian painter was subject to clear and even ironclad rules and his work had to serve the specific propagandist and religious goals of the regime, and art was usually created in order to aggrandize the king or the dead to both men and the gods. The painting and its message were supposed to be understood by the contemporary audience and were constructed in a way that enabled the meaning of the artist to be easily apparent. The representation of a man's movements were all known and recognized by the people to whom the art was directed. "The aim of the painter was to communicate. This is art in which each person could choose what was essential for the portrayal of the figure, to disregard any naturalistic visual representation as long as the object was recognizable. In practice it created a language of forms. In time these chosen forms became a canon" (Schäfer 1986, 150).

One of the fundamental factors that dictated the method in which the artists worked was the lack of perspective. There is usually no consideration paid to the notion of depth in the paintings. Any attempt at perspective that does exist is through the portrayal of one object partially overlapping another object, indicating that one object is in front and the other is behind. This may be seen in the picture in which a group of people is running and the legs and arms of those in front partially overlap those of the figures behind them (**fig. 3**). Another example can be seen in the portrayal of a chair being carried by two men (**fig. 4**). In it, the hand of the arm holding the chair from behind is actually drawn holding the front, with no consideration given to the depth of the chair being carried. In the case of the group of running men, we can tell that some of them are in the back and others in front but we cannot know how they are spread out in the space.

In the representation of the body itself, this lack of perspective resulted in a sort of flattening or spreading of the body; the depiction of the body is an amalgamation of specific parts drawn from the angle by which they are easiest to depict, resulting in a body composed of parts seen from different angles. The body is usually shown with the legs in profile, the furthest one stepping forward so that both legs can be seen. The line of the buttocks is clearly seen while the pelvis is rotated three-quarters towards us. The torso is also rotated in the same way, almost completely facing the front while one breast, or nipple in the case of a man, is drawn in profile. The entire width of the shoulders is represented frontally and the arms also face the front. The head appears in profile (Lexova 1935, 16-17). In practice, a person could not be in the exact position portrayed in most of the pictures. Since the lines flatten the form in a way that makes it easiest to represent each part, the direction in the picture does not necessarily indicate the direction in reality. When discussing how to understand a pose (or a movement), this principle is of great significance. A clear example of can be seen in the picture of a man sleeping on a small bed (**fig. 5**). The man is sleeping and his legs appear to be pointing upwards and his head, seen in profile, rests on its back part. In fact, "the customary sleeping position, according to three-dimensional representation, is on the side with the head resting on the ear on a head-support, so that the knees are facing us and not vertical as they are drawn in the picture. For the Egyptians, the sleeping position was known and they understood it according to this unrealistic picture" (Schäfer 1986, 251). This example makes it clear that in other cases, even in those where the body is turned or

Figure 5. Man sleeping on a bunk, New Kingdom, Berlin 20488 (after Shäfer 1986, 123).

Figure 6. Musicians-dancers, New Kingdom, Thebes Tomb of Nakht (after Shäfer 1986, 210).

Figure 7. King and queen in receiving offerings, Ancient Egyptian drawing from a statue of the same period, tomb, El Amarna, New Kingdom (after Shäfer 1986, 172).

rotated in a manner which seems realistic - as in the pictures of the musicians and the dancers (**fig. 6**) - one should still question the intentions of the artist in his portrayal.

An additional point when contemplating representations of Egyptian dance is the organization of space. In Egyptian paintings there is no reference to space as such. The figures can be placed anywhere and objects can be dispersed in the picture without any relationship to their real position in space[2]. Spatial connections occur through the creation of groups of figures or layers of figures without there being any change in their individual forms. This shows that there are multiple figures but does not explain their exact arrangement in space. Thus, for example, in a scene in which a royal couple is standing and receiving gifts, the queen seems to be standing behind the king (**fig. 7**). This is an Egyptian picture of an extant statue; in the statue, the royal couple is standing side by side, shoulder to shoulder. The Egyptians, who recognized the scene, knew this, but we would not have known this without the statue (Schäfer 1986, 172).

The division into registers appears in the Old Kingdom period (ca. 2500 BCE) when figures on a line, representing the ground, emerged. However, this still does not indicate the way the figures were dispersed in space and in which direction they were facing. This is undoubtedly true when the figures do not overlap each other (in a practical arrangement) as they do in the picture of the dancers from the Old Kingdom (**fig. 8**). Accordingly we cannot tell if they are positioned one behind the other or side by side, if the group of singers and dancers are turning in different directions, or if they are arranged in a semicircle, etc. There are many ways in which the spatial distribution could be understood and consequently the possible directions of movement are many and varied; we cannot know which is the 'correct' possibility nor what the artist intended. The Egyptian, however, saw these dances and recognized the scene from their own lives and knew the intentions of the artist.

To sum up, the Egyptian artists do not provide us with an accurate portrayal of the position of the body or its location

---

[2] For perception of space and time, see Frankfurt 1962.

Figure 8. Dance from the Old Kingdom, three singers, and three dancers, Giza, tomb 90 (after Shäfer 1986, 173).

in space in relation to other figures or objects. Consequently, we lack too much information to know with any certainty the actual arrangement of the figures to which the painter was alluding.

Even when I take the liberty of identifying these figures as dancers and the scene as dance, it is not possible to infer much about the way these dancers moved or the way the dance was performed as a whole. It is possible to derive a few hints from certain cases in which there are details that are emphasized or that are unusual[3]. There is nothing in them which provides a picture of one distinct possibility of movement but they do open the door to many possibilities that can be derived from them.

Characteristics of Ancient Egyptian Dance scenes

In light of the definition of dance presented above, the following discussion will examine the scenes that contain moving figures that I have identified as dancers. Taking into account the points raised earlier, and the difficulties in defining what dance is and what it is not, the question becomes much more complex when we are faced with silent pictures alone, and when the codes and the intentions behind these pictures remain open to our interpretation without any first-hand guidance or explanations (Garfinkel 2000, 16). The text accompanying the picture is often not clear and does not provide clarifying information (Brunner-Traut 1958). In this section, I will investigate the question according to the definition of dance and the points raised above, using individual examples to illustrate them.

According to the definition discussed above, it is the intention and the context of the situation or the event which defines the act or the performance as dance. Without any testimony or clear information concerning the intentions behind each scene, the possibility of classification is, in many cases, left to us. As a result, we can formulate either the most comprehensive or the most selective definition of what could be considered dance. The most comprehensive definition would include every movement or physical activity that does not portray any other action, and every group organization that does not involve any other categorized activity. According to this definition, people who are not trained dancers by occupation may be executing a dance as part of their ongoing activity, for example priests, worshippers or sometimes mourners. I will describe the examples which I present and the principles underlying them. If the narrowest definition is chosen, only those people that are organized symmetrically and deliberately into groups will be considered to be dancers. In many cases they are accompanied by the presence of musicians. Their position in the picture indicates a clear purpose in the execution of the movement and its character. For the most part, the dancers in those scenes will be performing identical or similar movements. Sometimes many different movements can be seen, performed by dancers who are identical in size and dress. An additional aspect that would determine dance within this narrow definition would be the character and type of the movement. Whether the movements are simple but specific, or so complex that they are acrobatic, in both cases the group clearly appears to be composed of trained and organized dancers, each one of whom is performing his role precisely in a position that was determined in advance. I will present assorted examples of these cases in this discussion.

Dancers do not engage in any other identifiable work

The first criterion for dance defined above is that the figures are not engaged in any clear, recognizable or identifiable work. Therefore, although the bending forms of the slaughterer or the kneeling bodies of the fishermen at work seem to be engaged in a complex and stylized movement, we do not consider them to be dancers. For example:

A slaughterer at a funeral feast, the tomb of Ninetjer, Old Kingdom (fig. 1)

In this picture a male figure on the left grasps an ox. The man is standing on his right foot and his entire body is leaning slightly forward diagonally. His left foot is bent

---
[3] For example the *Muu* dancers, who frequently appear to be touching the ground with the toes of the rear leg, and with the heel of this leg raised. This seems to indicate a dynamic, forward movement or even the beginning of a leap (**fig. 12**). In another example, several dancers have a round disk at the end of their braided hair, with the braid waving in a motion that may suggest a swinging, circular movement (**fig. 19**).

and lifted forward at a right angle to the supporting leg, while his back is inclined in a straight line that continues in the diagonal direction dictated by the supporting leg. The shoulder girdle is almost completely facing the viewer, thus twisted in relation to the lower part of the body, whereas his arms are extended at different heights in the same direction as his head and body, towards the right side of the picture. If the slaughterer was not holding a knife in one hand and the leg of an ox in the other, we could see an extremely dexterous dancer depicted in the middle of a dynamic movement. Since this movement does not exist separate from the context of its purpose - the subjugation and the slaughter of the ox - I do not regard it as movement that exists for its own sake. The emphasis here is not on the movement itself but on the successful fulfilment of the movement in slaughtering the ox. Therefore, although the pose of the slaughterers in this scene portrays dynamic and arresting body movements, I do not see them as dancing figures according to my definition. The figures that I regard as dancers are not engaged in any work but only in movement itself (**figs. 2, 8**). Sometimes these figures appear to be holding a musical instrument with which to accompany the dance and can also be seen playing instruments while dancing (figs. **10, 11**). In another case they are holding a stick that is used to hunt birds (**fig. 11**). Here it is clear that the tool held in their hand is not being used for its original function but as an object which plays a part in their dance (Spencer 2003, 114). Usually, however, the dancing figures do not hold anything in their hands and their arms are part of the entire body movement.

Figures that raise questions regarding this criterion are representations of priests (hieroglyph 26A in Gardiner) or worshippers (hieroglyph 4A) (**figs. 1, 9, 12**). It is difficult to clearly describe what these figures are doing and the purpose of their actions. At the same time, although their actions do not seem to have an intelligible and apparent goal, we see a stylized and specific movement of their bodies which can be portrayed with great attention to detail, such as the position of the fingers, the direction of the head and the bending of the legs. These figures, whose occupation is clearly sacred, might be considered dancing figures according to the present definition if dance serves as a means for carrying out their sacred work: prayer, ritual or any other ceremony (Scott 1997, 8). Even though I would not call them dancers *per se*, dance exists and is part of their functions as priests or worshippers. If we wished to incorporate only those for whom dance is a vocation, their occupation and their expertise as representatives of dance, we could limit the definition according to these criteria and in this case priests and worshippers would not be included.

The funerary priest at the tomb of Ninetjer, Old Kingdom (fig. 1)

The figure on the right in the top register is a funerary priest. He is standing to the right of the slaughtering scene and his body and the movement of his hand are inclined to the left, in the direction of the scene nearest to him and to the owner

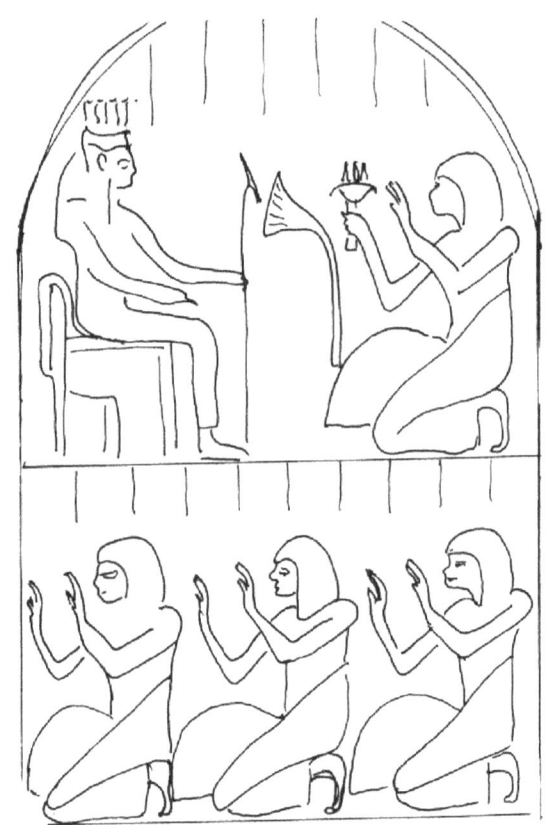

Figure 9. Worshipping the goddess Hathor, stele of the workman Nefersut, Deir el Medina, New Kingdom, (after Robins 1997, 188, fig. 223).

Figure 10. Dancer playing instrument and two musicians. New Kingdom, Thebes. Tomb of Nakht (after Kanawati 2001, 103).

Figure 11. Dancers, dwarf and musicians at a funeral feast. Tomb of Ninetjer Giza, Old Kingdom (after Junker 1951, fig. 4).

Figure 12. Muu dancers progressing towards the funerary priest during a funeral. New Kingdom Thebes, tomb of Tetiki (after Lexova 1935, fig. 59).

of the tomb. According to my first definition, the priest is not performing any physical activity with his body that would produce a visible and obvious result. Because of the nature of his role, he is performing some kind of ceremony. When we look at the position of his body, he is executing a movement in which his right hand is stretched out and bent forward slightly at shoulder-height. His hand and fingers are also fashioned in a deliberate and stylized manner. His left arm, with its clenched fist, is placed along his side. There is no doubt that his motions are purposeful and planned and all the elements of his body are precisely placed. The movement is not a complex one, but its very existence indicates intention by its execution. In my opinion, the movement in this case can be regarded as a dance within a ceremony. The movement here is a tool that represents, symbolizes and expresses the content of the ceremony (or a part of it).[4]

---

[4] In his article, Scott (1997, 8) cites Augustine's attitude to ritual dance. The citation explains that even if the ceremony itself has another purpose or objective, when the movement is emphasized in any way it can be said to be a dance within the ceremonial framework.

A dance scene incorporates a number of participants and not one individual dancer

An Egyptian dance scene mostly incorporates a number of dancers, two or more. The dancers do not appear as individuals except in a very few cases (Meyer-Dietrich 2009).

1. Ostracons that were found in the workmen's village of Deir el Medina, where a few records of individual dancers were found (**fig. 13**), may have served as the artist's sketchbook and been created from observations of community life. It is highly likely that they are an indication of spontaneous dancing that existed in the life of the community.
2. Another representation of an individual dancer which also seems to have the character of a sketchbook is seen in the picture of a dancer who appears in the apparent

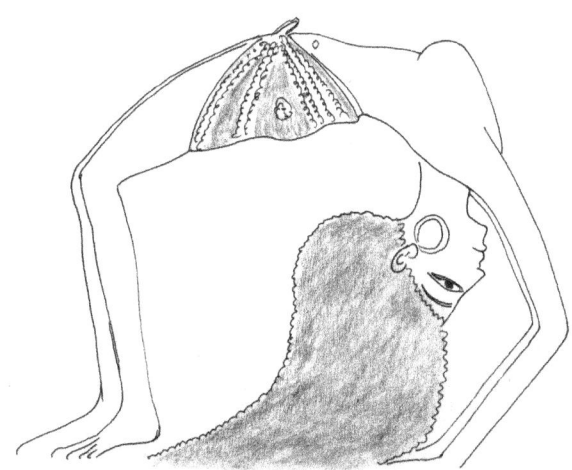

Figure 13. Girl in back-bend (bridge). Ostrakon Torino, 7053, Deir el Medina, New Kingdom (after www. Egyptyoga.com/umages68.jpg).

Figure 14. Girl in different stages of a movement, Beni Hasan, Middle Kingdom (after Lexova1935, fig.31).

progression of one movement along a timeline (**fig. 14**). This representation shows the development of a movement executed by the same dancer at different stages of the process, from standing to jumping (Lexova 1935, 19. Meyer-Ditrich 2009).

3. Single dancers only appear in pictures from the New Kingdom, when an individual dancer appears between a group of female musicians who are holding instruments and playing. It is usually the costumes which differentiate the dancers and the musicians (**fig. 10**) (Lexova 1935, 44).
4. Another representation of an individual dancer is that of the dancing god, *Bes*. This figure also appears in three-dimensional form (**fig. 15**) (Spencer 2003, 112).

In all other instances which appear in tombs or temples throughout all periods, the dancers are in groups. Here are a number of examples of a variety of dance scenes in which the dancers appear at least in pairs.

Dancers at a funeral feast. The western wall in the tomb of Ninetjer, Giza, Old Kingdom (fig. 11)

There are seven large female dancers in this picture and one dwarf. Three women are sitting in front of the dancers with their legs folded beneath them. They are clapping and looking at the dancers. The dancers are in two groups and while the pose of their bodies is slightly different it is clear that they belong to the same scene and are participating in a single event (Junker 1951, 127). Their costumes are similar; their movements are stylized and were coordinated in advance.

Muu dancers at a funeral ceremony, New Kingdom (fig. 12)

Three men are seen on the right, advancing in a uniform movement in the direction of the funerary priest standing to the left. The three men are wearing short skirts and on their heads are tall, narrow papyrus hats. Their hands are placed at their sides and their fingers are arranged in a distinctive position. Their legs are apart in a wide stride and the heels of their left legs are raised while the front

Figure 15. The dancing figure Bes, combining characteristics of a man and a lion. British Museum 20865 (after Spencer 2003, 112).

legs are in the air, creating a sense of a dynamic movement at the moment of transition (this is a transient position that cannot be sustained). This group of men appear only in the context of funerals and always at least in pairs (Brunner-Traut 1958, 53. Meyer-Ditrich 2009). The coordination of their movements indicates a prior choreography.

Dancers in a temple, festival of Opet to the god Amun-Ra, New Kingdom (fig. 16)

A group of male or female dancers are seen in a dynamic, acrobatic movement. These dancers are participating in the Opet ceremony which is celebrated in honour of the god *Amun-Ra* (Spencer 2003, 112). In this ceremony, a procession accompanies the god who is taken out from his temple and led to the temple of his wife *Mut*, and intricate acrobatic movements are seen. There is no doubt that the dancers who performed this dance were trained both as individuals and as a group.

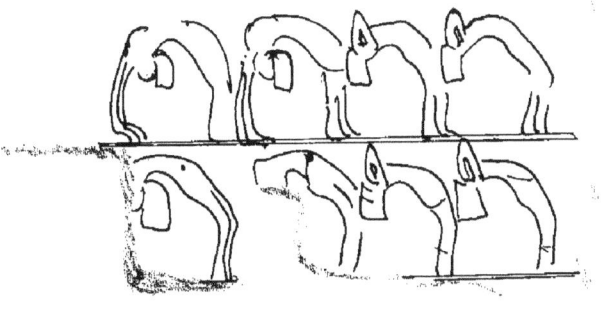

Figure 16. Male or female dancers in a back bend. Opet. Temple of Amon Ra, Luxor (after Spencer 2003, 112).

Dancers at the tomb of Kheruef, Thebes, New Kingdom (fig. 2)

The pair of dancers seen in the picture are part of a group dancing in a line. They are dancing as part of the jubilee festival (the Sed ceremony) in honour of the king (Strouhal 1997, 46). Their movements and costumes are identical. The movement of their bodies is complex as it involves twisting their spines and bending downwards at the same time. Their glance is directed upwards so that their necks are also twisted at an acute angle. Their arms and hands are also beautifully and precisely styled. The positions of the body and the obvious coordination between the dancers are evidence of their training and the precise planning of the dance.

Dancers at a feast, Tomb of Nebamun. Thebes, New Kingdom (fig. 17)

Two female dancers are seen in the picture, accompanied by musicians who are playing and singing during a feast (Aldred 1986, 176). The dancers appear to be almost naked, wearing only necklaces and a narrow band around their hips. They are intertwined in a joint dance whilst clapping. Even though their movements are not completely identical and their poses are slightly different from each other, there is no doubt that the dance is composed for both of them together and their movements complement one another.

Mourners, tomb of Ankhmahor, Saqqara, Old Kingdom (fig. 18)

The men and women in the picture are mourners (Kanawati 2001, 5). A physical, group event is taking place. Mourning is characteristically expressed by repeated movements, the hands touching the head or raised up to different heights, the body crouched in a sitting position. In this picture the figures lean so far back that they are almost falling. The mourning is taking place in a group and two or three people can be seen holding each other in a joint movement.

Figure 17. Dancers at a feast, New Kingdom, tomb of Nebamun, Thebes (after Spencer 2003, 113).

Figure 18. Men and women mourning, Old Kingdom tomb of Ankhmahor, Saqqara (after Kanawati 2001, 29).

In all these scenes and others like them, the dancers appear in a group. This can be understood as a socio-cultural statement regarding the nature of dance in Egypt. Dance takes place in a social context and deals with the connection between both man and man, and man and god, in festivals and in the context of death; between one dancer and another and between the dancers and the spectators.

Symmetry in the depiction of dance scenes

In our eyes there is beauty, perfection and appeal in symmetry. "Symmetry, as wide or narrow as you may define its meaning, is one idea by which man has tried throughout the ages to comprehend and create order, beauty and perfection" (Weyl 1965, 5). Symmetry is a practical concept in our everyday lives that we identify intuitively (Abas and Salman 1995, 32). We live symmetry in the arrangement of our bodies and in the arrangement of the environment in which we live. Symmetry is one of the primary elements that give us a sense of order, organization and harmony. "In the everyday sense, we use the concept of symmetry to describe balance and total agreement in size and form of opposite sides of a structure. In the wider sense it is attributed to organized patterns that are created by objects constructed from identical or similar units. The concept also relates to harmony of proportions (Abas and Salman 1995, 32). The definitions of symmetry in science are different and much more complex. I will treat symmetry here in the simplest and most accessible manner since our intuitive or instinctive attraction to symmetry is one aspect of its strength. It is a concept of form that is easy to understand intuitively and gives a sense of orientation and order. Pictures of many Egyptian dance scenes have a characteristically symmetrical structure, which is possible because they are composed of a number of figures that are usually performing the same or similar movements, either parallel to or opposite to each other. This arrangement draws the viewer's eye and creates a basic sense of identification and partnership with what is taking place. Two examples from the images presented above will be further discussed to highlight this structure; one is from the Old Kingdom, where there is a clearly symmetrical arrangement of the figures, side by side and opposite each other. This is a typical, although complex, example of many dance scenes. The second example depicts a pair of dancers from the New Kingdom in which another sort of symmetry can be seen and the pose of the dancers is much looser. Nonetheless, the symmetrical organization of the picture is easy to perceive.

Dance scene from the tomb of Ninetjer, Giza, Old Kingdom (fig. 11)

Order and symmetry can be seen in the structure of the line of figures – despite the difference between the two groups, their arrangement shows their structural connection and the clear relationship between them. The three figures on the right are parallel to the three on the left, and the middle group relates equally to the two groups at each end. This group can itself be divided symmetrically in the middle. The space between the middle group and the two outer groups is equal on both sides where, on one side, the small figure of the dwarf is seen, and on the other the two limbs of the dancer-musician. In the two spaces, the upper part of the register is empty above the tops of the dancers' heads. On closer scrutiny, differences between the two groups of dancers may be perceived. There are variations in their costumes and their movements. The left arms of the dancers on the right are bent and held above their heads with their forefingers and their thumbs touching each other (possibly producing a sound). The dancers in the middle group are holding sticks in their left hands along the side of their bodies[5], and in the hands of their right arms, which are bent and held high, they are holding sistrums (rattle-like musical instruments that are also associated with the worship of the goddess *Hathor* (Junker 1951, 135)). As for the structure of the picture, the figures in each group are duplicated, so that each one is clear and defined in itself. This duplication gives a general sense of order, direction and meaning, although scrutiny of the details shows us that there are slight differences in the figures. The differences can be seen in the dress, the position and size of the limbs, and even in the portrayal of the faces. The symmetry is not rigid but can be perceived in general, giving us a simple and immediate sense of the group, the directions and the movements within it. This is a picture that has a clear sense of organization, order and orientation, but at the same time it is not rigid or static but conveys a feeling of movement, change and dynamism.

A pair of dancers, tomb of Nebamun, Thebes, New Kingdom (fig. 17)

The two figures dancing together are also in a recognizably symmetrical arrangement. Here too, even though the dancers are not performing the exact same movements, they intersect one another, creating a harmonious structure that is symmetrically balanced. The movements of their hands, where the hands of the dancer at the rear are clapping downwards and the clapping hands of the dancer in front are held aloft, together creating a single diagonal line. In the same way, the lines of their hips combine to form ahorizontal line despite the slightly different angles at which they are bending. If we view them from a short distance, the two dancers look like one figure with double limbs intertwined. In the picture it seems that the central two legs of the dancers create the basis of their joint pose.

The representation of dancers in a shifting movement

Another less frequent arrangement of dancers is the representation of similar figures, exactly alike in dress and size, appearing one beside the other along a single line, in a similar movement that changes from one figure to another or from one pair of figures to the next. This image may indicate a large group of dancers simultaneously performing a dance that incorporates a variety of movements, or

---

[5] The Egyptians used sticks of this kind to hunt birds. This may be a hint that this was a "hunting dance".

alternatively the documentation of the development of the movements of a dance along a timeline, where the artist has decided to record specific movements of the dance. Although there is no apparent symmetrical structure in the representation of the scene, the figures appearing in it are in fact identical and may even represent the same figure over and over. Close observation of the entire mural gives a clear sense of dynamism and movement precisely because of the lack of rigid organization of the figures. Looking at one register separately shows us the possible development of the dance through the transition from one movement to another and from one pose to another, clearly revealing dynamic movement. This representation may be seen as a type of "movement notation", or could be read as notes for the performance of a dance or of a certain movement exercise.

Five registers showing dancers. Tomb of Mereruka, Old Kingdom (fig. 19)

On this relief dancers are arranged in five separate registers. The direction in which this picture should generally be read is from left to right, as all the figures on the left are turning their heads to the right as if they are advancing in that direction, and all the figures on the right are turning to the right as if they are also continuing to advance in the same direction[6].

When studying each register separately, two dancers clapping can be seen on the left of the highest register, on their right are a pair of dancers in skirts with a disc-like circle at the end of their plaited hair that seems to indicate that its weight has caused it to swing slightly away from their bodies. There are three pairs of dancers of this type, all of whom are clearly symmetrical and perfectly coordinated and each pair is holding hands in a joint pose that is different from the two other pairs. This series of figures can be read as three pairs of dancers dancing simultaneously in the same place, but recorded at three different points of the same dance. Alternatively, it is possible that they are perf+orming a complex and diverse dance that incorporates different movements and steps carried out at the same time. We can also see this picture as a record of the progression of different stages of a dance performed by the same pair of dancers. If we choose this reading, we can look at each register separately and read the progress of either a pair of dancers (in the two upper registers and in the lowest one) or a single dancer in each of the remaining registers. According to this reading, it is possible that these registers, in which the dancers are dancing separately, show the part of the dance in which they have separated (after dancing together in the two higher registers), later returning and meeting again for a joint dance in the lowest register. In view of the Egyptians' preference for an orderly and organized presentation of many dance scenes, I am inclined to read this picture as a depiction of the course of the movements of

Figure 19. Dancers, Old Kingdom, tomb of Mereruka, Saqqara (after Van Lepp 1985, 386).

one or two dancers and not a record of a dance with many participants and a diverse assortment of movements being performed at the same time.

A girl in different stages of movement. Beni Hassan, Middle Kingdom (fig. 14)

The girl in this picture is changing the position of her body from one picture to the next in a way that could portray different moments of one extended movement (Lexova 1935, 44). We have here a depiction of a movement that includes the transition from standing on one foot to standing on two, to a leap from both feet at the same time. There is no record here of the jump's landing. The course of the arm-movements is also clearly shown. It is hard to imagine that the intention here is to record several figures in different positions (even though this is possible), therefore I perceive this to be a 'page of instructions' for the execution of an exercise or the progression of movements of a dance. Alternatively, it may be an attempt by the artist to capture and record the characteristics of a movement.

Conclusion

How do we define dance in ancient Egypt? What are the

---

[6] An interpretation of the way in which the dancers are depicted and the relationship of this whole scene to the representation of words in hieroglyphic writing can be found in an article by Van-Lepp (1985).

tools and criteria according to which we can call one drawing a dance scene whereas another drawing will not be included in this definition? These are the questions that have been discussed in this paper.

The more I attempted to create the tools to identify dance and to formulate a clear definition of it, the more the complexity of the problem became obvious. Since I have been investigating this subject as a dancer, I have not been content to classify dancers and dance scenes merely on the basis of images. My trained eye notices movement and analyzes body postures almost as a regular habit, and this has brought me to question and study this subject, if only because so many of the representations of human figures in Egyptian art seemed to me to represent dancers. In the limited Egyptological literature that deals directly with this field, this question has not been raised explicitly, and the treatment of dance in Egypt deals mainly with categories, roles and different representations. There is no clear definition of the categories according to which a specific figure should be defined as a dancer as opposed to another figure that should not.

In order to find the necessary tools, I investigated the subject via contemporary dance and became aware of the complexity of the problems of definition. Since dance plays many roles in human society, defining it requires various tools. It must be defined as an art, as a social skill, as the means for social or personal expression for the individual, as part of religious ceremonies, etc. These are all different areas that have to be assessed in different ways. The role of the classical ballet dancer on stage is different to that of the folk dancer in the street or the ballroom dancer on the dance floor. In order to achieve a definition that can serve all of these, one needs to discover what it was that defined a certain set of movements as dance. I found that it was not possible to arrive at such a definition on the basis of movement alone, since many different sequences of movements could be considered as dance. However, certain dances would not be included within this definition on the basis of movement sequences alone. This shows that the human context, the environment and the intentions of both performer and spectator are the determining factors required to define a sequence of movements as a dance as opposed to any other physical activity.

When approaching this problem concerning the finds from ancient Egypt, I considered them only on the basis of what could be seen in the drawings. Since there is also a problem of definition regarding contemporary dance, how much more difficult must it be when we only have images to study? When we take into account the fact that the Egyptian artist worked according to a set of rules and portrayed these scenes in a language that was understood by his audience, rather than trying to portray a realistic picture, we are left with more questions than answers.

Despite all the issues and complexities mentioned, I have tried to examine the different types of representations and the definitions that I believe classify them as possible dance scenes. The guidelines were: examination of the shape of the pose, the form in which the scene is presented, the purpose of the activity, and the context in which it exists. All these provided both the tools for studying the different representations of movement and a method for determining whether they represent dance or are some other form of activity. They also exposed borderline cases where the existence of dance is open to question.

For me, this primary attempt to create guidelines for the definition of dance in ancient Egypt is the first step in an examination of many instances that intrigue both the eye and the mind concerning the character, nature and purpose of the movements and dance of this rich culture. Our most basic understanding of movement shows us that we cannot assume that we know how the dances portrayed in paintings and drawings were actually performed. This fact, to me, has two facets. On one hand, there is some sadness that I will never be able to see, experience or perform the portrayed dances in the way that they were lived and experienced in Egypt. On the other hand, because of my love of dance as an evanescent and unique art form, these pictures leave hints that are full of mystery and which offer many possibilities for creating dances inspired by them. Going deeper into the analysis, classification, and understanding of the many possibilities of movement concealed in the moving figures, we are given the opportunity to read these dancing figures as footsteps or clues that will enable us to bring them to life in many different ways.

References

Abas, S. J. and Salman, A. S. 1995. *Symmetries of Islamic Geometrical Patterns.* Singapore, World Scientific.

Aldred, C. 1986. *Egyptian art in the Days of the Pharaohs 3100-320 BC.* London, Thames and Hudson.

Alter, J. B. 1996. *Dance-Based Dance Theory, From Borrowed Models Lo Dance-Based Experience.* New York, Peter Lang Publishing.

Brunner-Traut, E. 1958, *Der Tanz im Alten Agypten.* Glückstadt, J. J. Augustin.

Dominicus, B. 1994. *Gesten und Gebärden in Darstellungen des Alten und Mittleren Reiches.* Studien zur Archäologie und Geschichte Altägyptens, 10. Heidelberg, Heidelberger Orientverlag.

Gardiner, A.H. 1957. *Egyptian Grammar: Being an Introduction to the Study of Hieroglyphs.* Oxford, Griffith Institute .

Groenewegen-Frankfort, H. A. 1972, *Arrest and Movement. An Essay on Space and Time in the representational Art of the ancient Near East.* New York, Hacker Art Books, Inc.

Green, L. 1983. Egyptian Words for Dancers and Dancing. In J. K. Hoffmeier and E. S. Meltzer (eds.) *Egyptological Miscellanies. A Tribute to Professor Ronald J. Williams. The Ancient World* 6(1-4), 29-38.

Garfinkel, Y. 2000. *Dancing at the Down of Agriculture.* Austin, University of Texas Press.

Helck, W. and Otto, E. 1975. *Lexikon Der Aegyptologie*. Wiesbaden, O. Harrassowitz.

Junker, H. 1951. *Grabungen auf dem Friedhof des Alten Reiches bei den Pyramuden von Giza (Giza X)*. Vienna, Rudolf M. Rohrer.

Kanawati, N. 2001. *The Tomb and Beyond*. Warminster, Aris & Phillips.

Lexova, I. 1935. *Ancient Egyptian Dances*. Prague, Oriental Institute.

Meyer-Dietrich, E. 2009. Dance. In *UCLA Encyclopedia of Egyptology*. University of California, CA (http://escholarship.org/uc/item/5142h0d)

Montet, P. 1925. *Les scènes de la vie privée dans les tombeaux égyptiens de l'Ancien Empire*. Paris, Strasbourg University.

McFee, G. 1995. *Understanding Dance*. London, Routledge.

Robins, G. 1997. *The Art of Ancient Egypt*. Cambridge, Harvard University Press.

Schäfer, H. 1986. *Principles of Egyptian Art*. Oxford:,Griffith Institute.

Scott, G. 1997. Banes and Carroll on Defining Dance. *Dance Research Journal* 29(1), 7-22.

Smith, W. S. 1981. *The Art and Architecture of Ancient Egypt*. Revised with Additions by William Kelly Simpson. London: Penguin Books.

Spencer, P. 2003. Dance in Ancient Egypt. *Near Eastern Archaeology* 66(3), 111-121.

Strouhal, E. 1997. *Life of the Ancient Egyptians*. Liverpool, Liverpool University Press.

Tefnin, R. 1984. 'Discours et iconicité dans l'art égyptien', *Göttinger Miszellen* 79, 55-72.

Tefnin, R. 1991. Éléments pour une sémiologie de l'image égyptienne. *Cronique d'Égytpe* 66, 60-88.

Van Lepp, J. 1985. The role of Dance in Funerary Ritual in the Old Kingdom. In: S. Schoske (ed), *Akten des vierten Internationalen Ägyptologen-Kongresses München 1985, Band 3: Linguistik - Philologie – Religion*, 385-394. Munchen, Helmut Buske Verlag.

Vandier, J. 1964. *Manuel d'archéologie égyptienne. Tome IV. Bas-reliefs et peintures. Scènes de la vie quotidienne*. Paris, Éditions A. et J. Picard et Cie.

Weyl, H. 1965. *Symmetry*. Princeton, Princeton University Press.

Wild, H. 1956. *La danse dans l'Égypte ancienne. Les documents figures*. Positions des thèses des élèves de l'École du Louvre (1911-1944). Paris, Ecole du Louvre.

Dance Dating in the Old Kingdom; Formal Rules, Step 1: Know thy Dances.
Establishing a Typology of Old Kingdom Dance

Lesley J. Kinney

Abstract: The dating of Old Kingdom tombs has been the subject of much scholarship. This paper presents a working typology for dance styles in Old Kingdom Egypt and how changing trends may be used to further date some tombs within the Old Kingdom.

There are a number of fundamentals to consider when establishing a typology of dance genres for a culture existing in the remote past. First of all, the relevant primary evidence at hand comprises visual pictures of dance, or what we perceive to be dance, and some accompanying captions used to describe the scene contents as well as titles of performers.

Further, we must consider whether the images depict dance genres or different dance steps which could be employed in a number of different dance genres.

Finally, how do we label genres? Some styles have terms applied to them traditionally by previous scholars which may be labels according to the form of dance or the context in which it is performed, while other genres have labels which have been attributed to assumed origins or associations which may be misleading.

Introduction

This paper examines the considerations encountered in attempting a classification of ancient Egyptian dance. It is also about how such a classification can be used as a tool for the dating of Old Kingdom tombs and, in particular, it is an opportunity to present my Typology of Dance for the Old Kingdom.[1]

There are a number of fundamentals to consider when establishing a typology of dance genres for a culture which existed in the remote past. First of all, the material evidence of dance in ancient Egypt comprises pictures of dance, or what we perceive to be dance, as well as some accompanying captions used to describe the scene contents, titles embedded in the scene and the biographies of individuals who were part of the performance cohort.

In determining which scenes would be subjected to my methodology, I included all scenes that had previously been considered to be representations of dance by others, whether or not I agreed, and scenes which had not been included by others if they complied with my basic definition of dance as rhythmic movement. My full definition of dance is published in the introduction to my book (Kinney 2008, 1-3).

As much as this study is aimed at the establishment of a typology of dance genres, the possibility must also be considered that we are merely classifying the various motifs employed by artists in representing dance, whether or not they reflected the actual nature of dance performed at the time. My opinion is that it is most probable that the forms represented resemble the forms performed, as observed by the artists, which were appropriate to the context depicted.

As some ancient Egyptian scenes can be analysed as enlarged determinatives or hieroglyphs, it is possible that some may not represent dance as it was performed, but rather capture it in its most representative form. Nevertheless, determinatives used to differentiate dance terms are reasonably varied and it is probable that there is a resemblance between the representation, even in writing, and the actual dance forms performed.

Further, we must consider whether the images depict dance genres or different dance steps which could be employed in a number of different dance genres and contexts. In some instances, only one pose or step is depicted and we are faced with the dilemma of whether to categorise according to pose, or to deliberate as to whether an entire dance form is represented by this one characteristic step.

Perhaps a major reason for discrepancies in the way ancient Egyptian dance has previously been classified is due to the way the Egyptians themselves classified dance. In most cases, we have only the evidence of isolated poses, as represented in wall art, whereas the ancient Egyptians knew the dance forms in their entirety and within the framework of the context in which they were performed. Similarly,

---

[1] There is no unanimous agreement amongst Egyptologists regarding numerical dating relative to our system of dating for the Old Kingdom. Many accept it was in the 3rd millennium BC but there is compelling evidence for an earlier dating. For a detailed hypothesis as to dating options for Old Kingdom dynasties see Cherpion 1999 (which attempts to date tombs to specific reigns in the Old Kingdom) or Kanawati 2003 (a compelling and interesting hypothesis for Old Kingdom dating based on the cattle count and utilizing forensic DNA analysis).

some modern authors, such as Brunner-Traut (1958), Arroyo (2003), and Anderson (1995), have attempted to classify according to dance genre, whilst others have classified according to the pose represented. For example, Wild (1963) was particularly thorough in identifying poses, but still categorized according to genres based on overall dance style and context. Dominicus (1994) made the most consistent attempt at categorizing according to pose alone, regardless of other criteria, but her thesis is not exclusively or comprehensively an examination of dance. Because of the nature of certain dance forms, even when categorized according to pose, the resulting classification may in some instances describe a whole dance genre rather than an isolated step or pose. This exposes a flaw in the classification process according to the criterion of form.

I could find no solution to this dichotomy. Therefore, in instances where only one pose is depicted, the style or genre classifications in this typology may, in the final analysis, be no more than a categorization of specific dance poses. These poses may have been used in more than one dance form in more than one context, in much the same way that a modern dance step such as the *pirouette* can be performed in classical ballet, musical comedy, tap, folk, or even ballroom dance genres, all with different context, purpose and audience. It follows that a typology based primarily on the form of the pose represented allows for the most consistent classification. Yet some categories include varied poses and can be more readily identified by the grouping of dancers as an aspect of form.

Furthermore, how do we label genres? Some styles have terms applied to them by previous scholars, which may relate to the form of the dance, or the context in which it is performed, or the performers themselves, while other genres have been assigned labels attributed to their assumed origins or associations, which may be misleading.

Because we are not able to witness ancient Egyptian dance first hand, we can only classify representations of dance according to the criteria apparent in them, such as:

1. Provenance - dating and location of examples
2. Context and function
3. Costume
4. Props
5. Type of accompaniment
6. Labels, titles or identifying captions
7. Grouping of dancers
8. Form of pose
9. Performer (i.e. gender)
10. However, some of these criteria proved to be limited as a starting point in classifying the scenes.

Dating

Using this criterion was only possible where the origin of the example was known. While the provenance has been established for most of the scenes in this study, the dating of tombs in Old Kingdom Egypt has been the subject of many studies, and various means have been employed to establish a system for placing tombs accurately into specific reigns within the Old Kingdom. Thus, the dating of scenes is not used to influence the categorization process in this typology, as it was recognised that the classification of dance genres may in itself become a viable tool for the dating of tombs, and this has proven to be the case. For example, tracking dance genre trends chronologically became a useful dating tool; and the position of the arms in the Diamond Dance proved to be very specific to dating tombs within the Old Kingdom.

Context and function

While considered important criteria in the classification process, context and function were not the primary consideration for classification in this study because, for a significant proportion of dance scenes, the context is damaged, absent or simply not published. Further, various dance forms appear in more than one context and a number of apparently different dance forms appear together in quite a few scenes. For example, the Diamond and Salute Dances appear in sacred, secular, and funerary contexts.

Costume, props

Sometimes the evidence is too fragmentary to identify the costume specifically. However, the criterion of props and paraphernalia is especially relevant to the classification of certain genres and the formation of sub-groups within others.

Type of accompaniment

Accompaniment can be melodic, that is, with orchestral accompaniment, or purely rhythmic, accompanied by rhythmists who clap the pulse. Again, the type of accompaniment is often not apparent due to the fragmentary nature of the examples. In this study, accompaniment did not appear to be a significant differentiating criterion between genres.

Label or identifying caption

There are a number of Egyptian terms used to describe dance embedded within dance scenes, some of which indicate differentiation of purpose and context. For example, the term *h3t* occurs only in funerary scenes, but appears with more than one dance form. This suggests that the ancient Egyptians themselves may have had a concept of classification based on entirely different criteria to those employed by modern authors. Ancient systems of classification are not necessarily based on criteria or methodology similar to our own, as illustrated by M. Foucault (1970, preface). Ancient Egyptian terms do not appear to lend themselves to a consistent system of dance classification as numerous apparently different forms of dance often share the same terminology. This is especially true of the dance term *ib3*, which appears as a caption with no less than four distinctly different genres and in more than one context.

Performer/ form of pose

It became apparent during the course of this study that the criteria present in most scenes would be the best starting point for classification; that is, the information that could be gleaned from examining the form of the dance and the grouping and gender of the dancers.

Out of well over 100 pictures of dance, form of pose could be discerned in all but three cases: fragments from the Solar Temple of *Š3ḥw-R'* (Borchhardt 1981-82, pl. 54), and in the tombs of *Nfr-m3't* (LD II, 17(c)) and *K3.i-ḫnt* (el-Khouli and Kanawati 1990, pl. 40). Once pictures were restricted to the general time frame of the Old Kingdom, form proved to be the best choice for the classification of dance scenes into genres. When arranged into genre groups, much of the analysis would consist of tracking variations and developments in the form and depiction of dance throughout the history of the Old Kingdom and across various regions.

Grouping of dancers, form of pose

Whilst most dances could readily be classified by form, others - such as the Pair Dance which has numerous postures - were more readily identifiable by the grouping of the dancers. It therefore became apparent that the grouping of dancers is an equally important consideration to the form of pose depicted.

Typologies based on the grouping of dancers have been established for the classification of dance pictures in studies of ancient dance in other cultures. In a rare example of a formal typology of dance scenes, Malaiya (1989, fig. 17.1) uses the grouping of dancers as the principle criterion for classifying dance scenes in the rock art of Central India. Dance scenes are first grouped according to whether they are performed by solo dancers, duos, or groups, and then the scenes are further classified according to form, accessories, context, and function. Only Decker (1994, 753ff) primarily categorizes Old Kingdom dance scenes according to the grouping of dancers rather than pose alone, using such labels as *Formationstänze* for all unison dances including Diamond, Salute, Swastika and Layout Poses. However, he does not use labels to differentiate each pose. It has proved difficult, in practice, to reconcile classification according to pose consistently with a system of classification according to grouping.

Dance groupings could thus be broken down into the following categories:

1. Unison Dances (i.e. simultaneous execution of the same dance step), such as Diamond, Salute, Linked Hand and Layout. Poses such as the Swastika would have to be split, because some examples are unison while others have multiple styles in unison, some have multiple direction, and others show one step as a sequence of movements.
2. Pair Dances. Some examples feature unison pairs and others show different poses; some are static while others are dynamic; and some examples are performed by women while others are performed by men.
3. Interactive Group dances. These include genres such as the Boomerang Dance and Sistrum Dance, and interactive tableaux such as the Mirror Dance and Boys' Game. The Boys' Game genre is classified as dance in this study because the Egyptian dance term *ḥbt* appears in the caption accompanying the example on the British Museum fragment (James 1961, pl. 25). In this case, more than one motif or grouping of the performers is used to identify the genre even though they are not all present in each example, which in turn makes this a dance genre rather than a classification of pose alone.
4. Solo Dances such as the scene featuring a dwarf dancing in the tomb of *K3.i-'pr(w)* (Fischer 1959, figs. 17 and 18).
5. Multiple styles performed in the same register, including *Enchaînements* in which the steps included are sometimes represented in sequence. Thus, while this genre comprises a series of forms classified according to pose, the overall group represents a single dance genre.
6. Yet another configuration is notation style which records an entire dance sequence. The most comprehensive example can be observed in the tomb chapel of *W'tt-ḥt-ḥr* inside the mastaba of *Mrrw-k3.i* (Kanawati and Abder-Raziq 2008, pl. 60).

The final prioritization of criteria used for the classification process is as follows:

1. Dating: general dating to the Old Kingdom only
2. Performer: form of Pose; grouping of dancers; gender
3. Criteria used to classify specific genres and sub groups: context and function; props and costume; caption; accompaniment; dance style versus dance pose.

Pictorial Overview of genres (refer to figures)

Once a system of classification had been established and examples placed into their genre groupings, the next consideration was the application of suitable names for each of the categories. In some cases, this was straightforward since previous authors had already furnished names that have been widely adopted and still appear appropriate for the corresponding categories. As far as possible, these already well-established genre names have been adopted in this typology. When a suitable term had not already been established, a name has been applied which primarily describes the form of the dance. Following is a description of each dance genre as classified in this study with, where necessary, a brief explanation of the criteria used to establish the classification and references to the labels assigned by previous authors. The poses and dance genres

Figure 1. Dance Style 1, Tomb of Idw, Giza (after Simpson 1976, fig. 38).

Figure 2. Dance Style 2a, Tomb of Queen Mr.z-ꜥnḫ III, Giza (after Dunham-Simpson 1974, fig. 11).

are presented here in the chronological order of their earliest known appearance in the Old Kingdom. For a more detailed analysis of each dance style refer to Kinney (2008, 53-167).

Dances appearing from the 4th Dynasty

Genre 1 – Diamond Dance (fig. 1)

The Diamond Dance is the most commonly depicted dance genre, appearing in both presentation and funerary contexts. Captioned *ib3* by the Egyptians, the Diamond-shaped arm position in this stately dance resembles both cows' horns and the symbol ⊔ for *k3*. Perhaps the most problematic genre to name, the label Diamond Dance has been adopted for Dance Style 1 in this study. Numerous writers have commented on the resemblance of the form of the pose to the bucranium: "Stately dances, which might be taken to symbolize the horns of a *Cow*" (Anderson 1995, 2563), "*la danse aux bras relevés, imitation ou symbole des cornes d'un bovidé*" (Hickmann 1954-5, 154), and "cow-dances" (Watterson 1996, 113). Other names include: "The Purely Movemental Dance" (Lexova 2000, 21), "offering table dance" (Meeks 2001, 357), "*Strenger ib3-Tanz*" (Brunner-Traut 1958, 15), "dancing with upraised arms" (Saleh 1998, 483) '*Rauten-Schreittanz*' (Brunner-Traut 1975, 219), '*la station immobile*' (Wild 1963, 38), "the more austere style" (Sameh 1964, 124), and "Dance of *iba*" in which the dancers' arms form "a rhombus shape" (Arroyo 2003, 343). Only Watterson (1996, 113) has actually named it "cow-dance"). The decision to apply the Diamond Dance nomenclature was based on the geometrical configuration of the arm position rather than the assumption that the dance was necessarily associated with the cow-goddess Hathor. While there are indications for this, the dance also has connections with other cults and beliefs.

Genre 2 – Salute (figs. 2-5)

Dance Style 2 was categorized previously by Wild (1963, 38), who named it *Salute à la romaine*. In this study, it is named simply Salute as there are numerous variants, some of which do not resemble the salute associated with Rome.

Figure 3. Dance style 2b, Tomb of Ḥm-mnw, el-Hawawish (Kanawati 1980-1989, V, fig. 6.).

Figure 4. Dance Style 2c, Tomb of Ni-m3ˁt-Rˁ, Giza (after Roth 1995, fig. 188).

Figure 5. Dance Style 2d, Tomb of K3-gm-n.i, Saqqara (after Decker and Herb 1994, S 3.51, pl. 421).

The Salute genre has been considered a variant of Dance Style 1 by Brunner-Traut, Arroyo, and Anderson, and it has been omitted by Meeks, Saleh, and Lexova, perhaps due to the same assumption. While the two forms appear in the same contexts, presumably with the same purpose, the categorization according to form was adhered to here for the sake of consistency, and because form is an indicator of the nature of the dance, the study of which was a primary objective of this study.

Sub-group 2a (fig. 2)

Also labelled *ib3* by the Ancient Egyptians, this dance has a processional appearance. The forward arm is held up in

Figure 6. Dance Style 3a1, Tomb of Queen Mr.z-ꜥnḫ III, Giza (after Dunham-Simpson 1974, fig. 11).

front of the face and the elbow is bent, resembling a salute or greeting, hence the label of Salute in this study. This genre also appears in both funerary and presentation scenes.

Sub-group 2b – straight arm with flexed hand (fig. 3)

A variation of 2a, this pose has the forward arm rigidly held straight out in front at varying heights, with the hand flexed back.

Sub-group 2c — ḥn gesture (fig. 4)

The arm position in this Sub-Group is the same as that in Sub-Group 2a, but the wrist is bent and the hand held parallel to the ground. A number of examples have the label ḥn embedded in the scene.

Sub-group 2d (fig. 5)

The front arm is lowered but the back arm is bent acutely and the hand placed against the temple or behind the ear.

Genre 3 – Swastika Dance and Dances with batons (figs. 6-10)

Dance Style 3 is classified according to form and then divided into sub-groups according to the various batons employed by the dancers, although there is no certainty that these sub-groups are associated. A number of writers have considered the Dance Style 3 sub-groups as one and assigned general labels, which are in most cases more appropriate to the variants performed with weapons than those variants executed without weapons.

Previous labels include: "The war Dance" (Lexova 2000, 30); "war or combat dance" (Saleh 1998, 483); "so-called boomerang dancers of purported Libyan origin" (Saleh 1998, 483); and "Libyan type dance" (Saleh 1998, 483). The various forms of this dance have been previously distinguished by Brunner-Traut as "*Schwingtanz*" (for the Swastika Pose) (Brunner-Traut 1958, 219), and both "*Jagd Tanz*" (Brunner-Traut 1975, 219) and "*Grüppentanz mit Hölzern*" (Brunner-Traut 1975, 29) for similar dances performed with batons, as well as "*ein Tanz nach Libyscher Art*" (Decker and Herb *1994, I*, 711, 715-716). Lexova (2000, 44-45) takes the gazelle-headed throw sticks of some of these variants to be "probably a rattle" and includes them in her category 'Dance with Musical Instruments' (Lexova 2000, 44). Arroyo also divides this genre into groups: a variation of the 'Dance of Iba' for the Swastika Pose, particularly when performed with the Diamond Pose (2003, 348), 'The Sistrum Dance' given only to examples of dance with sistra (2003, 361), and 'Clapper Dance' for examples with clappers (2003, 364).

Sub-group 3a1 –Unison Swastika Pose (fig. 6)

The arrangement of the dancers' limbs in this genre resembles the shape of a swastika, hence it is referred to as Swastika Pose in this study.

Variants appearing in the 5th Dynasty

Sub-group 3a2 –Solo Swastika Pose **(fig. 7)**

In this cluster, a soloist in the Swastika Pose performs with a chorus of dancers executing the Diamond Pose.

Sub-group 3b - Boomerang Dance **(fig. 8)**

This sub-group is called Boomerang Dance because some or all of the dancers hold boomerang-shaped throwing sticks in one or both hands (although dancers hold zoomorphic throwing sticks in some examples). The boomerang-shaped and the zoomorphic throwing sticks are hunting implements which appear to have originated in Southern France in the MagdalenianPeriod (Garcia-Galloway-Lommel 1969, 251 and figs. 23, 363). It is possibly a variant of Dance Style 3a.

Figure 7. Dance Style 3a2, Tomb of K3.i-m-ˁnḥ (G 4561), Giza (Kanawati 2001, pl. 35).

Figure 8. Dance Style 3b, Tomb of K3.i-ḫnt, Hammamiya (el-Khouli and Kanawati 1990, Pl. 67).

Sub-group 3c – Sistrum Dance (fig. 9)

The pose in Sub-Group 3c is a less dynamic variation of the one represented in Sub-Group 3b, with one or more dancers holding a sistrum in one hand.

Sub-group 3d – Static poses with batons (fig. 10)

Sub-Group 3d includes examples in which figures in a variety of static poses hold various batons. The Swastika form is not apparent in any of these examples.

Genres appearing from the 5th Dynasty

Genre 4 – Harvest Dance (figs. 11-13)

The Harvest Dance appears to represent field workers executing a rhythmic movement performed in unison as part of the harvest. This genre has been omitted as a dance category by some authors, perhaps because they do not consider it a dance form. It is included as a genre here because it has previously been classified as dance and because the running and clapping of sticks in unison, at least in some examples, suggests rhythmic movement -

Figure 9. Dance Style 3c, Tomb of Nw-ntr, Giza (after Junker 1951, pl. 44).

Figure 10. Dance style 3d, Solar temple of Ni-wzr-R^c, Abusir (Borchhardt 1907, pl. 16:274).

which is the essence of dance at its most fundamental level. Previous authors have employed the following labels: Vandier "*Danses spontanées a la fin de la moissons*" (1964, 415); Meeks, "*rwi* – runaway dance" (2001, 356); Arroyo "Race Dance" (2003, 363); Lexova "Dance at Harvest Time" (2000, 23); Saleh "agrarian work" (1998, 482) and "stick dances at the harvest" (Saleh 1998, 482, 483). The label "*Erntetanz des Alten Reiches*" is used by Decker and Herb (1994 I, 840-841) but only a small sample of scenes which correspond to those classified in Sub-group 4b and 4c in this study are cited.

The word *rwi* appears in the Pyramid Texts (PT 884 (a) and 743 (d)) with stick-wielding figures as a determinative (Junker 1940, 1-39), and consequently has been associated with this and the Boomerang Dance. Yet there do not appear to be any Old Kingdom dance scenes captioned with the word *rwi* and, since this dance only appears in harvest scenes, it is named after the context in which it appears rather than the form of the pose or the batons held by the performers. The Sub-groups within this paper are classifications according to pose, interaction of dancers and type of baton.

Sub-group 4a – Rhythmic movement with long sticks (fig. 11)

In Sub-Group 4a, the men hold long batons with no evidence of percussion.

Sub-group 4b – Rhythmic movement and percussion with short sticks (fig. 12)

In Sub-Group 4b, the sticks are held in both hands and are struck together while the men run in unison, suggesting rhythm and dance.

Sub-group 4c – Combat with sticks (fig. 13)

In Sub-Group 4c, two men confront each other in a mock combat suggestive of a combat-style dance.

Genre 5 – Pair Dance (fig. 14)

This dance has many poses but is always executed in pairs. It appears in both funerary and presentation scenes. A small number of examples are accompanied by the caption *mk trf itt* … ["*behold the movement of…*"].

There has been almost unanimous agreement regarding the label 'Pair Dance' for Dance Style 5: "The Pair Dance" (Lexova 2000, 26); "*Reigenartige Paartanz*" (Brunner-Traut 1958, 21) and "*Paartanz*" (Brunner-Traut 1975, 219); "pair dances" (Saleh 1998, 483); "paired dancers" (van

Figure 11. Dance style 4a, Tomb of Ty-mry, Giza (after Weeks 1994, fig. 39).

Figure 12. Dance Style 4b, Nfr-b3w-Pth, Giza (after Weeks 1994, fig. 9).

Figure 13. Dance Style 4c, K3i-m-nfrt, Saqqara (after Simpson 1992, fig. F).

Lepp 1985, 392); "*Paartänze*" (Decker and Herb 1994 I, 737 ff); and "Pair Dance" (Arroyo 2003, 356). Only Meeks has differentiated naming it *trf* dance (Meeks 2001, 356), adopting the Egyptian term associated with dances in pairs. While the word *trf* appears only in captions accompanying this genre, it is present in only three out of 15 possible Old Kingdom Pair Dance representations of the genre.

Genre 6 – Enchaînements; steps in sequence including Pirouette, High Step, Stride, Kick, Linked Hand Dances and other miscellaneous steps (figs. 15-21)

A number of distinct poses are grouped together in this category because they are represented in sequence, therefore indicating a true dance genre.

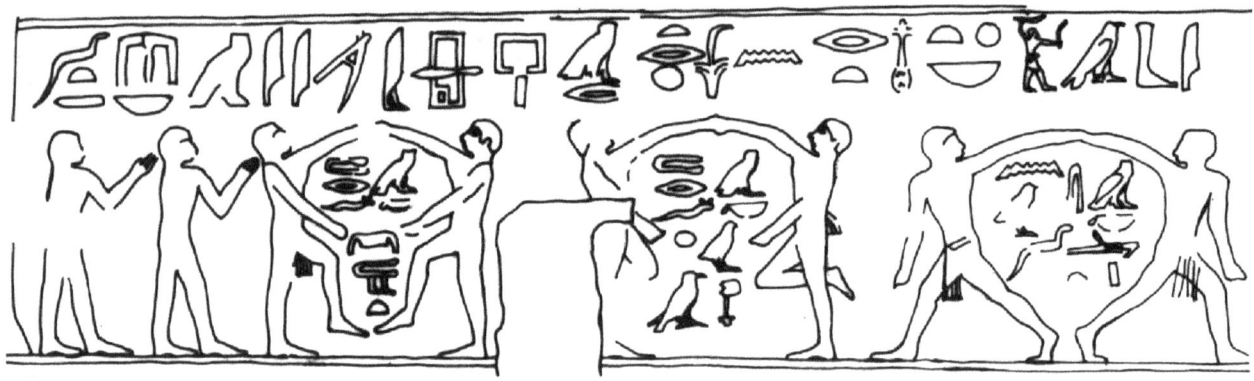

Figure 14. Dance Style 5, Tomb of Iy-mry, Giza (after Weeks 1994, fig. 35).

Sub-group 6a – Pirouette *(figs. 15 and 16)*

Sub-Group 6a, the Pirouette, has previously been classified for the Middle Kingdom as "Pirouette dancing" (Lindsay 1963, 138); and "the Pirouette" (Saleh 1998, 483; Wilkinson 1854/1998, 505; Wild 1963, 88). Henri Wild further differentiates two distinct variations as *"pirouette simple et pirouette fouettée"* (1963, 38). Depicted in the tombs of W'tt-ḥt-ḥr, Thy and perhaps Mrrw-k3.i, this dance step appears in three forms in the Old Kingdom: the Solo Pirouette, the *Fouetté* Pirouette, and the Partnered Pirouette, the latter with examples representing the pair both facing each other and back to back. The Partnered Pirouette has been examined as part of the Pair Dance genre by numerous other authors, but has not been identified as a turning step. The one Old Kingdom example which may indicate the *Fouetté* Pirouette (from the tomb of Thy at el-Khokha) is also considered as part of the Kick Step (see below 6c, fig. 19).

6a1 – Solo Pirouette *(fig. 15)*

In this genre, the weight is on one leg, with the other leg bent and the foot or ankle held against the knee of the supporting leg. The pose resembles the most characteristic moment in the execution of a pirouette.

6a2 – Partnered Pirouette *(fig. 16)*

The Partnered Pirouette also appears in a back to back form, and is also examined with the Pair Dance genre.

6b High Step *(figs. 17-18)*

Like the Pirouette, the High Step also appears in solo and partnered form.

6b1 – Solo High Step *(fig. 17)*

In this pose, the working leg is lifted high and bent at the knee like a marching step.

Figure 15. Dance Style 6a1, Tomb of Mrrw-k3.I Saqqara, (after Duell 1938, pl. 164).

6b2 – Partnered High Step *(fig. 18)*

Previously, the Partnered Pirouette and the High Step have only been considered as part of the Pair Dance genre, but they are considered here as a possibly distinct genre.

Sub-group 6c – Kick Step *(fig. 19)*

The example of the Kick Step from the tomb of Thy could also be considered as a *Fouetté* Pirouette. Vandier describes this movement as "goose stepping" (Vandier 1964, 418).

Sub-group 6d – Stride *(fig. 20)*

The Stride genre has been omitted by previous writers, with the exception of Saleh who identifies the scene from the tomb of Wnis-'nḥ as dance in the tomb report: "The dancers Stride slowly..." (Saleh 1977, 14). This genre appears as

Figure 16. Dance Style 6a2, Tomb of Mrrw-k3.I, Saqqara (after Duell 1938, pl. 87).

Figure 17. Dance Style 6b1, W'tt-ḫt-ḥr Chapel, Mastaba of Mrrw-k3.i, Saqqara (after Roth 1992, fig. 10, drawing by Mary Hartley).

a preparation for other steps in most examples; however, one register is dedicated to it in the tomb of *Wnis-'nḫ* at el-Khokha (fig. 20; Saleh, pl. 3)

Sub-group 6e – Linked Hand Dances (fig. 21)

This chorus dance, in which the dancers link hands, was well represented in the Pre-dynastic cave art of Upper Egypt (Winkler 1938, pl. 24:2), although the only Old Kingdom examples known to exist are found in the tomb of *'nḫ-ty-fy* at el-Moalla. Previously, the genre has been identified as dance by Vandier, but he does not suggest a name for it (Vandier 1964, 418-422).

Other scenes with no parallel:

A number of scenes which appear to depict dance do not fall into any of the above categories, particularly those appearing in the *W'tt-ḫt-ḥr* chapel, which are depicted in sequence with other steps examined in this category. Various poses in the *W'tt-ḫt-ḥr* chapel have been identified by Roth as a birthing dance sequence (Roth 1992, 141-143.)

Genre 7 – Funerary Dance (figs. 22-23)

Sub-group 7a. Dance of the Mww *(fig. 22).*

The Dance of the *Mww* was performed as part of funerary ritual. Most depictions of *mww* dancers occur after the Old Kingdom - the word *mww* only appears in one Old Kingdom scene, in the tomb of *Nb-k3w-ḥr*. The dance term used in captions for *mww* dance is *ḫbi*, but does not occur in this context until after the Old Kingdom. Because the dance is identified chiefly by the characters who perform it and their distinctive costume, previous labels for this genre have tended to refer to the *mww* who perform this dance rather than the dance itself: "muu dancers" (Reeder 1995; 1, Meeks 2001, 357-58), and "Muu ritual dancers" (Anderson 1995, 2564). Even when the dance itself was named, labels such as "Dance of the *Mww*" or "*Mww* Dance" always referred back to the performers -"*Muu-Tanz*" (Brunner-Traut 1958, 43); "*Les danses des Mouou*" (Wild 1963, 37); "*La danse des Mouaou*" (Jequier 1927, 144); "*Der Tanz der Mww*" (Junker 1940, 1; Decker and Herb 1994, I, 723

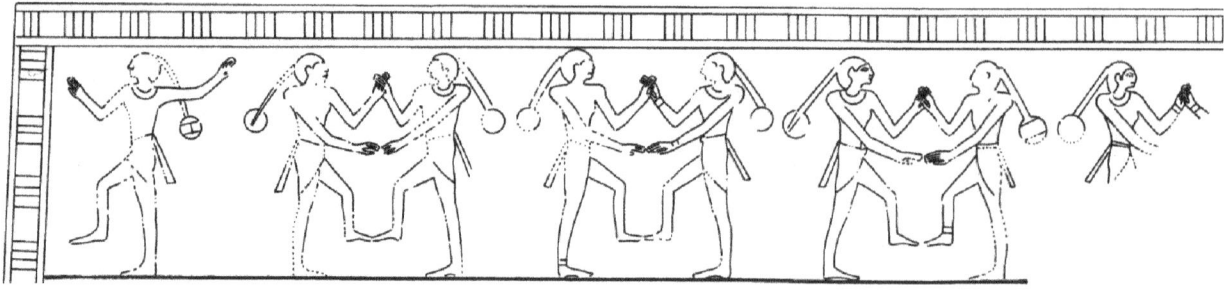

Figure 18. Dance Style 6b2, Tomb of Ibi, Deir el-Gebrawi (Kanawati 2007, pl. 69).

Figure 19. Dance Style 6c, Tomb of Ihy, el-Khokha (after Saleh 1977, pl. 18).

Figure 21. Dance Style 6e, Tomb of ʿnḫ-ty-fy, el-Moalla (after Vandier 1950, pl. 33).

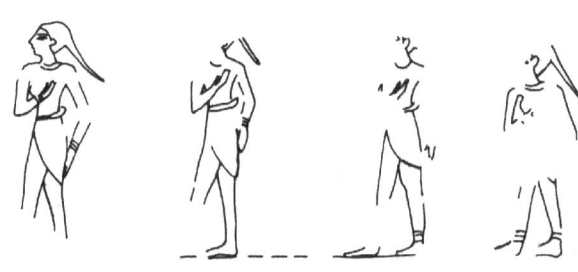

Figure 20. Dance Style 6d, Tomb of Wnis-ʿnḫ, el-Khokha, (after Saleh 1977, pl. 3).

ff); "The Dwarf Dances" (Lexova 2000, 62); and "Dance of Muu" (Arroyo 2003, 362). Even though the pose is consistent in the Old Kingdom examples, the classification results in a dance genre rather than a particular pose.

Sub-group 7b – (W)nwn Dance **(fig. 23)**.

The Old Kingdom (W)nwn funerary dance formerly identified only by Junker (1940, 8-9) has had little previous analysis. In the Old Kingdom, it appears in a funerary context only in the tomb of *Ptḥ-ḥtp*. Since the actions and functions performed by the dancers, as described in the Pyramid Texts, are the same as those performed by the *mww*, the dance has been considered to be a possible variant of the *mww* genre. The posture identified with the *Wnwn* Dance is executed by female dancers in most examples occurring after the Old Kingdom, and in the hieroglyphs describing it in the Pyramid Texts. A detailed analysis of the differences between *mww* dance and *wnwn* dance is given in Kinney (forthcoming).

Genre 8 – Boys' Game *(fig. 24)*

Dance Style 8 has been given various names which were suggestive of the perceived intent behind the performance. These were "Post-circumcision initiation rites" (Saleh 1998, 482), and "foreigner or hut games" (Lindsay 1965, 121), while Decker and Herb describe the two main motifs as "Ausländerspiele" and "Hüttenspiele" (1994 I, 623-625). The term "Boys' Game" was coined by Wilkinson (1854/1988, 100) and this term has been adopted here since it neither pre-suggests the purpose of the game or performance nor assumes that it is necessarily dance. The dance term *ḥbt*, appearing in the British Museum scene (EA 994; James 1961, pl. 25), suggests the genre was considered to be dance by the Egyptians themselves. Since this genre appears near a depiction of the Mirror Dance, and, because one participant carries a hand-shaped clapper similar to those held by dancers in the Mirror Dance scene, the impression given is that it is associated with dance.

Genres appearing in the 6th Dynasty.

Genre 9 – Layout Pose *(fig. 25)*

Previously, Dance Style 9 has been described as acrobatic or gymnastic dance - "acrobatic performances" (Anderson

Figure 22. Dance Style 7a, Tomb of Nb-k3w-ḥr, Saqqara, (after Junker 1940, fig. 1).

Figure 23. Dance Style 7b, Tomb of Ptḥ-ḥtp, Saqqara (after Lepsius 1901, Erganz 43).

Figure 24. Dance Style 8, Mastaba of Mrrw-k3.i, Saqqara (after Duell 1938, pl. 162).

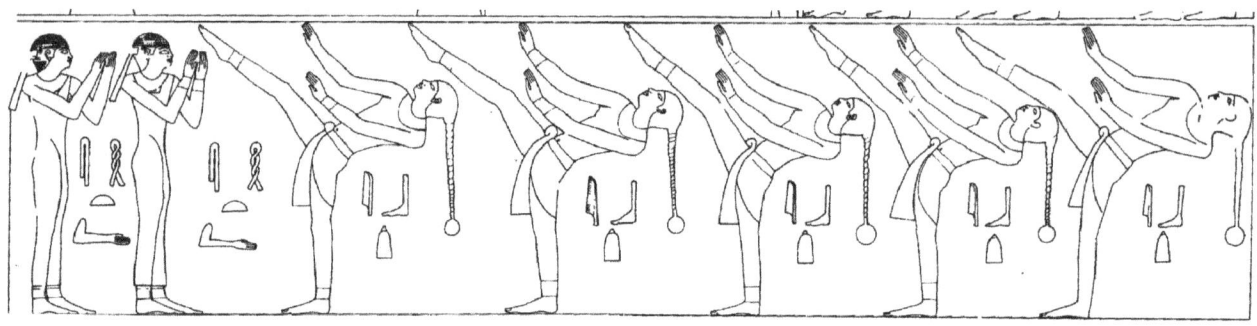

Figure 25. Dance style 9, dancers in the tomb of ꜥnḫ-m-ꜥ-ḥr, Saqqara (Kanawati-Hassan 1977, pl. 58.).

Figure 26. Dance style 10, tomb of Mrrw-k3.i, Saqqara (after Duell 1938, pl. 164).

1995, 2563); "acrobatic dance" (Saleh 1998, 483); and "The Gymnastic Dances" (Lexova 2000, 22). This is included with all unison dances in the general classification "*Formationstänze*" by Decker and Herb, (1994 I, 774 ff), and called "Dance of the Stars" by Arroyo (2003, 353). It is renamed here according to the modern dance term for the pose represented. While this pose is undeniably acrobatic in nature, the labels acrobatic or gymnastic are too broad and allow for confusion with other steps, more acrobatic in nature, which were introduced in the Middle Kingdom. In this case, naming the pose rather than the dance style rules out any confusion.

Genre 10 – Mirror Dance (fig. 26)

In this genre, the dancers hold mirrors and hand-shaped clappers. A chant or hymn to Hathor appears in the scene from the tomb of *Mrrw-k3.i*, suggesting the genre is associated with her cult. The dancers in both examples of this genre are arranged in an interactive rondo formation. The term Mirror Dance has been employed by all previous writers: "Mirror Dance" (Saleh 1998, 483); "mirror dance" (Anderson 1995, 2563; Saleh, 1977, 14); "*La danse aux miroirs*" (Hickmann 1954-5); "Spiegeltanz" (Brunner-Traut 1975, 219; Decker and Herb, 1994 I, 839-840); and "Dance of the Mirrors" (Arroyo 2003, 350). There are only two exceptions: "*Hathor Sprungtanz*" (Brunner-Traut 1958, 22, but see above 1975, 219) and "Children's Games" (Lindsay 1964, 118), in which it is considered together with examples of the Boys' Game genre in a classification of children's games. Classification of this genre according to the grouping of dancers and their props resulted in a categorization of it as a dance style rather than a pose.

Conclusion

The classification of Egyptian dance offers a number of problems, including whether to categorize according to pose, or complete dance form, or a combination of both as appropriate; whether some poses can be used to visually represent different dances in different contexts and with different purpose; or whether some poses/dances should be considered as belonging to more than one genre grouping. I hope this article has established a working model for a Typology of Dance styles in the Old Kingdom.

References

Anderson, R. D. 1995. Music and Dance in Pharaonic Egypt. In Sasson, Jack M. (ed.) *Civilizations of the Ancient Near East*, Vol. IV, 2555-2568. New York, Scribner.

Arroyo, R. P. 2003. *Music in the Age of the Pyramids*. Madrid, Natural Acoustic.

Borchhardt, L. 1907. *Das Grabdenkmal des Königs Ne-user-Re*. Leipzig, J.C. Hinrichs.

Borchhardt, L. 1981-82. *Das Grabdenkmal des Königs S'3hu-Re'*. 2 volumes. Osnabrück, O. Zeller.

Brunner-Traut, E. 1958. *Der Tanz im alten Ägypten*. Glückstadt, J. J. Augustin.

Brunner-Traut, E. 1975. Der Tanz. In *Lexikon der Ägyptologie* Vol. V, 215-231. Wiesbaden, Harrassowitz.

Cherpion, N. 1989. *Mastabas et hypogées d'Ancien Empire: Le probléme de la datation*. Paris, Connaissance de l'Égypte ancienne.

Decker, W. and Herb, M. 1994. *Bildatlas zum Sport im alten Ägypten Corpus der bildlichen Quellen zu Leibesübungen, Spiel, Jagd, Tanz und verwandten Themen*. 2 volumes. Leiden, Brill.

Dominicus, B. 1994. *Gesten und Gebärden in Darstellungen des Alten und Mittleres Reiches*. Studien zur Archäologie und Geschichte Altägyptens, 10. Heidelberg, Heidelberger Orientverlag.

Duell, P. 1938. *The Mastaba of Mereruka*. 2 volumes. Oriental Institute Publications 31. Chicago, University of Chicago Press

Dunham, D. and Simpson, W. K. 1974. *The Mastaba of Queen Mersyankh III. G75030-40*. Giza Mastabas 1. Boston, Museum of Fine Arts

el-Khouli, A. and Kanawati, N. 1990. *The Old Kingdom Tombs of el-Hammamiya*. Sydney, Australian Centre for Egyptology.

Fischer, H. G. 1959. A Scribe of the Army in a Saqqara Mastaba of the Early Fifth Dynasty. *Journal of Near Eastern Studies* 18, 233-272.

Foucault, M. 1970. *The Order of Things: An Archaeology of the Human Sciences*. New York, Pantheon.

Garcia, L. Pericot, Galloway, J. and Lomell, A. 1969. *Prehistoric and Primitive Art*. London, Thames and Hudson.

Gardiner, A. H. 1957. *Egyptian Grammar: Being an Introduction to the Study of Hieroglyphs*. Third Edition. Oxford, Griffith Institute.

Hickmann, H. 1954-55. La danse aux miroirs. *Bulletin de l'Institute Égyptien* 37, 151-90.

James, T.G.H. 1961. *Hieroglyphic Texts from Egyptian Stelae etc. in the British Museum, Part I*. London, British Museum.

Junker, H. 1951. *Giza X. Der Friedhof südlich der Cheopspyramide. Westteil*. Vienna, Rudolf M. Rohrer

Junker, H. 1940. Der Tanz der Mww und das Butische Begräbnis im Alten Reich. *Mitteilungen des Archäologischen Instituts für Ägyptischen Altertumskunde in Kairo* 9(1), 1-39.

Kanawati, N. and Hassan, A. 1970. *The Teti Cemetery at Saqqara, Vol II: The Tomb of Ankhmahor*. Sydney, Australian Centre for Egyptology.

Kanawati, N. 1980-1989. *The Rock Tombs of el-Hawawish: The Cemetery of Akhmim*, 9 Vols. Sydney, Australian Centre for Egyptology.

Kanawati, N. 2001. *Tombs at Giza, Volume I; Kaiemankh (G4561) and Seshemnefer I (G4940)*. Warminster, Aris and Phillips.

Kanawati, N. 2003. *Conspiracies in the Egyptian Palace: Unis to Pepy I*. London, Routledge.

Kanawati, N. 2007. *Deir el-Gebrawi Volume II, The Southern Cliff: The Tomb of Ibi and Others*. Oxford, Aris and Phillips.

Kanawati, N. and Abder-Raziq, M. 2008. *Mereruka and his Family Part II, The Tomb of Waatetkhethor*. Oxford, Aris and Phillips.

Kinney, L. J. 2008. *Dance, Dancers and the Performance Cohort in the Old Kingdom*. British Archaeological Reports Int. Series 1809. Oxford, BAR Publishing.

Kinney, L. J. Forthcoming (2013). The *wnwn* Funerary Dance. In Kousoulis P. and Lazaridis N (eds.) *Proceedings of the 10th International Congress of Egyptologists, Rhodes*. Orientalia Lovaniensia Analecta. Leuven, Peeters.

Lepsius, R. 1901. *Denkmäler aus Ägypten und Äthiopien*, 12 Vols. (I, II& III), Leipzig, Hinrichs (orig. 1849-59).

Lexova, I. 2000. *Ancient Egyptian Dances*. New York, Dover (reprint, original published, 1935).

Lindsay, J. 1965. *Leisure and Pleasure in Roman Egypt*. London, F. Muller.

Malaiya, S. 1989. Dance in the Rock Art of Central India. In Morphy, H (ed.) *Animals into Art*, 357-368. London, Taylor and Francis.

Meeks, D, 2001. Dance. In Redford, Donald, (ed.) *Oxford Encyclopedia of Ancient Egypt*, 356-360. Oxford, Oxford University Press.

Roth, A. M. 1995. *A Cemetery of Palace Attendants Including G2084-2099, G2230+2231 and G2240*. Giza Mastabas 6. Boston, Museum of Fine Arts.

Roth, A. M. 1992, The *pss-kf* knife and the opening of the mouth ceremony: a ritual of birth and rebirth. *Journal of Egyptian Archaeology*, 78, 113-47.

Saleh, M. 1998. Dance in Ancient Egypt. In Cohen, S. J. (ed.) *International Encyclopedia of Dance: A Project of Dance Perspectives Foundation*, 481-486. New York, Oxford University Press.

Saleh, M. 1977. *Three Old Kingdom Tombs at Thebes*. Mainz, Verlag P. von Zabern.

Sameh, Waley-el-dine. 1964. *Daily Life in Ancient Egypt*. New York, McGraw-Hill.

Simpson, W.K. 1992. *The Offering Chapel of Kayemnofret in the Museum of Fine Arts, Boston.* Boston, Museum of Fine Arts

Simpson, W.K. 1976. *The Mastabas of Qar and Idu. G 7101 and 7102.* Giza Mastabas 2. Boston, Museum of Fine Arts

Van Lepp, J. 1985. The role of Dance in Funerary Ritual in the Old Kingdom. In S. Schoske (ed), *Akten des vierten Internationalen Ägyptologen-Kongresses München 1985, Band 3: Linguistik - Philologie – Religion,* 385-394. München, Helmut Buske Verlag.

Vandier, J. 1964. *Manuel d'archéologie égyptienne. Tome IV. bas-reliefs et peintures. scènes de la vie quotidienne.* Paris, Éditions A. et J. Picard et Cie.

Vandier, J. 1950. *Mo'Alla. La tombe d'Ankhtifi et la tombe de Sébekhotep.* Bibliotheque d'Etude 18. Cairo, Imprimerie de l'Institut Français d'Archéologie Orientale.

Watterson, B. 1996. *Gods of Ancient Egypt.* Stroud, Sutton Publishing.

Weeks, K. R. 1994. *Mastabas of Cemetery G 6000. Including G 6010 (Neferbauptah); G 6020 (Iymery); G 6030 (Ity); G 6040 (Shepseskafankh).* Giza Mastabas 5. Boston, Museum of Fine Arts

Wild, H. 1963. Les danses sacrées de l'égypte ancienne. In *Les danses sacrées.* Sources Orientales VI, 68-69. Paris, Éditions du Seuil.

Wilkinson, J. G, Sir. 1988. *The Ancient Egyptians; A Popular Account.* New York, Bonanza Books (reprint: original edition 1854).

Winkler, H. 1938. *Rock Drawings of Southern Upper Egypt.* 2 Vols. London, The Egypt Exploration Society.

DANCE IN THE PREHISTORIC AEGEAN

Christina Aamodt

Abstract: The aim of this paper is to examine dance performances in the Prehistoric Aegean. Depictions of dancing appear in the Prehistoric Aegean as early as the Early Cycladic period. However, it is during the Minoan period that dancing becomes an important part of religious ceremonies, associated especially with the appearance of the deity, but also with rites of passage. Dance during the Mycenaean period seems to be mostly connected to funerary rites. Dancing, therefore, was considered by both the Minoans and the Mycenaeans an important part of religious ceremonies. In addition, dances could have been performed in secular occasions as well, suggested by some fresco representations, mostly from Crete. Nevertheless, dance is a very powerful means of expression, since it can assault all five senses simultaneously and convey meaning through many channels and at different levels. For this reason, the performance of dancing could also have a social and political function as a way to transmit complex social and political messages.

Defining Dance

Dance has been defined as human behaviour composed of purposeful, intentionally rhythmical and culturally patterned sequences of nonverbal body movements other than ordinary motor activities, where the motion has an inherent and aesthetic value (Hanna 1987, 19). Dance has also often been defined as "patterned movement" having an end in itself which usually, though not always, transcends utility (Spencer 1985, 1; Hanna 1987, 21). Dance is a multidimensional phenomenon directed towards the sensory modalities (Hanna, in Royce Peterson 1977, 197). What makes dance unique and what seems to make people choose it in specific instances over other means of expression is its kinaesthetic property and its ability to establish subliminal communication (Royce Peterson 1977, 194, 195). The kinaesthetic activity of dancing generates kinaesthetic responses in the viewers, where the capacity to assault all five senses simultaneously and to convey meaning through many channels and at different levels is perhaps what makes dance such a potent, and often threatening, vehicle of expression (Royce Peterson 1977, 194, 200). Ambiguity –which may be intentional- and the communication of conflicting messages are also possible due to the multi-channel nature of dance expression (Royce Peterson 1977, 200).

However, an examination of the theories concerning dance[1] shows that a definition of what dancing is in its essence is very difficult and often subjective, which indicates that its true nature still eludes us or simply that dance has many functions depending on the context and the occasion in which it takes place (Royce Peterson 1977, 208). This is further emphasized by the fact that dance is not a universally identical behavior. Hence in reality there can be no universal definition of what dance is (Hanna 1987, 30; McFee 1992, 49), or if there was it would be in some way lacking (German 2005, 68).

Therefore, dance has often been viewed as a means of social control (Hanna 1987, 134-142), or as a way to reestablish order after it has been disrupted (Brinson 1985, 208; Lonsdale 1993, 259), as a way of relieving excess energy which otherwise may turn on the society itself (Spencer 1985, 4; Hanna 1987, 142), as a way of unifying society and encouraging solidarity by creating feelings of strength (Spencer 1985, 14-15; Brinson 1985, 208) and as a vehicle through which identity is stated and maintained (Hanna 1987, 142-143; Maners 2006). Some studies have attributed a therapeutic value to dance, or an educational role (Spencer 1985, 3-11; Blacking 1985), whereas other anthropological studies have hinted at the element of competition displayed in dancing (Spencer 1985, 21; Hanna 1987, 136-137). Very often dance is associated with religion, as one way of worshipping or embodying the supernatural and/or as a means to achieve altered states of consciousness (Hanna 1987, 101-127, esp. 106, 133). Through dance people can communicate with the divine (Hanna 1987, 107) and this notion is usually emphasized in archaeological studies concerning dance, for example in Classical Greece (Burkert 1985, 103; Lonsdale 1993, 74-75). Finally, in order to understand how dance can convey messages, its communicative aspects have often been compared with that of a verbal language, assuming that dance functions in the same way language does (Hanna 1987, 86; contra Peterson 1977, 32).

Different societies define dance differently and consider different gestures as dance. Dance is a product of a specific peoples and their social system (Williams 2004, 35). It cannot be seen as something outside a specific culture and a specific social system, because it is shaped by it. In other words, dance as an action of the world is, at the same time, an interpretation of the world, because it is also shaped by it (Sklar, 2006, 106). Therefore, in order to understand what dance means to a specific society, it should be considered and examined as part of this society (Hanna 1987, 19). This can be particularly helpful when examining cultures and societies that have left no written evidence of what their dances were and the specific reasons and beliefs behind them.

---
[1] For a more detailed examination of the various theories concerning the anthropology of dance, see Williams 2004.

The evidence for Dancing in the Prehistoric Aegean

In regards to the prehistoric Aegean (c. 3100-1100 BC) an indirect indication for the performance of dancing is the representation and the existence of musical instruments, usually found in cult places and burial contexts. These instruments consist of stringed ones (lyres, harps, phorminxes), wind instruments (double reeded aulos) and triton shells used as trumpets and seistra (Younger 1988, 1-2). The earliest and most famous representations of such instruments in the Aegean are the Early Cycladic (EC) figurines depicting musicians (c. 3200-2300 BC), such as the seated harpist and the standing figure playing a double flute from the island of Keros (Liveri 2008, 3). These figurines are found in tombs, suggesting their significance in cult and could perhaps imply the performance of some ritual involving music and/or dancing in connection to funerary rites. In terms of the Mycenaean period actual specimens of musical instruments, such as tortoise shells interpreted as lyres dating to Late Helladic (LH) IIIA2-IIIC (c. 1370-1065 BC), were found in the sanctuary of Phylakopi on the island of Melos (Renfrew 1985, 325-326, 383-384), whereas ivory fragments of phorminxes were found in chamber tomb 81 at Mycenae and the Menidi tholos tomb on the mainland of Greece (Younger 1988, 10, 21, 22). Representations of figures playing musical instruments are more numerous, e.g. the famous blue monkeys from Xeste 3 at Akrotiri (Younger 1988, 15) on the island of Thera dated to Late Minoan (LM) I (c. 1600-1450 B.C), the lyre player from the palace of Pylos (LH IIIB2) and the figure depicted next to a giant lyre on a krater dating to late LH IIIA-early LH IIIB (c. 1370-1180 BC) from Nauplion on the mainland (Dragona-Latsoudi 1977, 89). Figures playing lyres were also found on Crete, specifically on the Ayia Triadha sarcophagus and the Ayia Triadha fresco, both dated to LH IIIA1 (c. 1390-1370 BC), and on a LH IIIB (c. 1310-1180 B.C) pyxis uncovered at Chania (Younger 1988, 23). The association of music with dancing is suggested by the depiction on the Harvester's vase (LM I) also from Ayia Triadha, of a procession of men singing and rattling a seistron. The fact that the figures have one leg up, indicating a vigorous rhythm, could suggest the performance of some kind of dance (Younger 1988, 2, 6, 8; Marinatos 1993, 137). Similarly, a clay model uncovered in a shrine at the settlement of Palaikastro (c. 1400 BC), also on Crete (**fig. 1**), depicts three female figures dancing to the sound of a lyre played by another female figure (Mandalaki 2004, 16).

The most important visual codes to indicate dance in the Prehistoric Aegean, according to Senta German, are arm gestures. These consist of: a) one arm up in front of the body and one arm behind, b) holding both arms in front, c) both hands held up at the sides, d) one hand up at the side, one hand down, e) both hands in the air, f) both arms out and down –always found in antithetical pairs like holding hands, and g) both arms down at the sides (German 2005, 56). The fact that the same gestures are also found in Minoan and Mycenaean processions renders the identification of dance scenes uncertain. According

Figure 1. Clay model of women dancing from Palaikastro, Crete (after the Archaeological Museum of Herakleion guide, p. 155).

to the same scholar the distinction between a dance and a procession in iconography is the fact that processions have an end point, whereas dances do not, in addition to the fact that in processions the figures are usually carrying objects (German 2005, 57). According to German there are several glyptic images that depict dancing (German 1999, 280; 2005, 56-68) and it is true that a sense of rhythmic movement is depicted in a number of them. However, this is a matter that deserves a more detailed examination, which exceeds the purpose of this study. Therefore, the emphasis will be on images that are generally believed to depict dancing.

The only representation so far of dancing from the Cyclades is found on a marble plaque at Korphi t'-Aroniou on the island of Naxos, dating to EC II-III (c. 2700-2000 BC). The scene depicts three figures shown in silhouette with raised arms and legs apart. The left figure is turned towards the other two and seems to be brandishing a club of some sort over the head of the middle figure (Doumas 1965, 57; Liveri 2008, 4). All three figures have been drawn on different levels, perhaps in order to depict a circular dance (Doumas 1965, 57). The difference in level, the raised arms and the bending of the knees of the figures indicate vigorous movement. However, whether the scene is a dance-scene or a hunting-scene is difficult to tell with certainty. It is worth mentioning, however, that a sealstone from the tholos tomb at Vapheio (CMS I.226) depicts a female figure in a hide skirt and bare breast holding a stick who seems to be dancing (LH II, c. 1500-1450 BC). The hide skirt indicates the religious status of the figure (Evans 1930, 70; Nilsson 1950, 155; Aamodt 2006, 125, 130, 161).

The evidence of dancing from Minoan Cretespans the Pre-palatial to the Post-palatial period (c. 3000-1100 BC) and consists of depictions in large and small-scale art. Dance in Minoan culture seems to have been closely associated with the epiphany of the deity (Evans 1930, 74; Lawler 1965, 32; Hägg 1983, 184), but also with initiation rites. Iconographical evidence suggests that the Minoans believed that the appearance of the divinity was possible through ecstatic dance (Warren 1988, 14; Marinatos 1993, 177; Liveri 2008, 5). In this case, dancing was most likely performed by a priestess impersonating the divinity and was a ritual action related to vegetation, tree and pillar cults and also to the cult of the dead (Branigan 1993; Liveri 2008, 4). Dance was apparently part of all three phases of the epiphany and was therefore performed in order to invoke the appearance of the deity, during the actual appearance and as an offering to the divinity after the latter's epiphany (Liveri 2008, 4). The epiphany could be either envisioned or staged. In the first case, the deity would appear as a vision to each worshipper, whereas in the second case a member of the royal family or a priestess would play the role of the deity (Niemeier 1987, esp. 165; Hägg 1986; Liveri 2008, 5). Such representations of dancing in the presence of or in order to induce the appearance of the deity are found on Minoan Crete as early as the Proto-palatial period (MMI-II, c. 2000-1700 BC). A circular pedestal-table and a bowl both uncovered in the palace of Phaistos depict three dancing female figures in bell skirts. In the case of the pedestal-table the central figure is holding up flowers (Goodison and Morris 1998, 121; Mandalaki 2004, 16; Liveri 2008, 5), whereas in the bowl the central figure is static and the other two seem to be dancing around her. Female dancers also seem to be depicted on the base of the pedestal-table. They are presented horizontally but they seem to convey the idea of a circular dance (Liveri 2008, 6)[2]. The gestures they perform are usually interpreted as dance, rather than worship (Marinatos 1993, 149; Goodison and Morris 1998, 123) and are, according to German's classification, pose E for the central figure of the pedestal-table and pose A for the other figures. The figures in the middle of the composition on both objects may represent a goddess (Marinatos 1993, 149; Schiering 1999, 748; Liveri 2008, 5), or alternatively may be priestesses with a leading role in the ritual performed (Mandalaki 2004, 16), though this notion seems less likely for the static figure on the bowl. Whichever their identity, they are clearly differentiated from the other figures and most likely represent either an ecstatic or a staged epiphany of the deity.

Some terracotta and bronze figures may similarly represent female dancers engaged in ecstatic dancing (Rethemiotakis 1998, 128), such as a bronze figure from Palaikastro dating to MM I (c. 2000-1850 BC) with its hands on its hips (Liveri 2008, 7, pl. 1,1). The latter is reminiscent of the female figures from the cult place of Ayia Irini (MMIII-LMIB/LHII, c. 1700-1400 BC) on the island of Keos (Caskey 1986, 34). The site has provided fragments of well over fifty statues, all of them upright (**fig. 2**). They all wear a full skirt and tight-waisted jacket with short sleeves and a

Figure 2. Clay statue from the temple at Ayia Irini, Keos (after Caskey 1986, pl. 4a).

girdle around the waist (Caskey 1986, 36), an outfit which is common both in Minoan and Mycenaean iconography. The hands of the statues rest at their waists and some of them wear a heavy ring around their neck, interpreted as a garland (Caskey 1986, 36). Similar terracotta figures of female dancers were also uncovered in the Neopalatial Shrine at the Minoan villa of Kannia/Gortys (Liveri 2008, 7, pl. 1,2), whereas the Palaikastro clay model mentioned before could also indicate ecstatic dancing since the figures have their arms extended but not touching, which would have allowed them to perform turns around themselves (Mandalaki 2004, 16).

During the Neopalatial period in Minoan Crete (c. 1700-1450 BC) depictions of dancing continue to be associated with the epiphany of the deity and/or with initiation rites. The iconography on gold rings is often related to ecstatic dances, involving a tree, a shield, a stone and other objects, which could similarly have an epiphany-conjuring purpose (Niemeier 1990, 168). The epiphany of the divinity is indicated by the presence in some scenes of flying birds, which in Minoan religious iconography are thought to indicate the physical presence of the divinity (Nilsson 1950, 330-340; Niemeier 1990, 168). These scenes usually depict a female figure in the Minoan flounced skirt in the middle of the composition. The figure is flanked by one or two figures, one of them male. The male figure is usually shaking a tree, which grows either from a cult structure, probably a shrine, or from a pithos, as on the gold ring from Vapheio on the mainland (CMS I.219), dating to LH II (c. 1500-1450 BC),

---

[2] For an examination of circular dance performances in the prehistoric Aegean, see Soar 2010.

most likely imported from Crete (Niemeier 1990, 169). The central female figure on the ring from tholos tomb A at Archanes (LM IIIA1) has been recognized as a goddess by a number of scholars (Sakellarakis 1967, 280; Niemeier 1990, 168), her gesture interpreted as that of epiphany rather than dancing, a notion that can be supported by the floating objects near her. These objects are also found in scenes depicting single figures interpreted as deities and therefore should be understood as symbols of the appearing goddess (Niemeier 1990, 168).

Nanno Marinatos has connected these depictions with initiation rites involving the psychological manipulation of the participants who experienced contrasting moods by handling the tree (Marinatos 1993, 187).The interpretation of the cult rites depicted in these scenes is indeed difficult, but it is possible that they included invocations, ecstatic dances or other mimetic acts connected to the epiphany and the worship of a deity, which would have been the culminating point of the rites (Liveri 2008, 11).

The most characteristic example of an invocation dance, however, is the gold ring from the Isopata tomb near Knossos (**fig. 3**). The Isopata ring (LM I-II, c. 1600-1450 BC) depicts three female adorants or priestesses dancing in a meadow indicated by the presence of lilies. Another female figure is placed in the middle and on a higher level, perhaps suggesting she is the high priestess (Marinatos 1993, 10), whereas the deity is represented as a diminutive hovering figure descending from the sky (Marinatos 1993, 177; Liveri 2008, 9 contra Niemeier 1990, 168). The high priestess has one arm down, the other bent and upraised, making the blessing gesture.

The most well-known representation of a dance is the so-called "Sacred Grove and Dance" fresco from Knossos (**fig. 4**), a depiction that led Sir Arthur Evans to outline the religious function of dance, describing it as orgiastic and ecstatic (Evans 1930, 68-70; German 1999, 279). The fresco depicts in the foreground at least fourteen female figures, most of whom have one arm raised up in front and the other held down in front or behind (German 1999, 279; Jacobs 2004, 10). The female figures are separated from the rest of the action by a border, believed to depict a paved causeway (Immerwahr 1983, 145; 1990, 65; Davis 1987, 157; Marinatos 1987a, 141; Jacobs 2004, 11; German 2005, 51). On the upper part of the fresco, on the other side of the causeway, two large separate groups of men and women are represented, and in the middle there are two or three large trees (Marinatos 1987a, 142; Jacobs 2004, 10), the eponymous Grove. The female figures in the foreground were initially interpreted to be dancing (Evans 1930, 67), a notion that was supported by other scholars (Mouratidis 1989, 51; Immerwahr 1990, 66; Kontorli-Papadopoulou 1996, 104). Some, however, have rejected this interpretation due to the fact that the hair of the figures does not fly in the air nor do their feet dangle (Davis 1987, 158; Marinatos 1987a, 141; Jacobs 2004, 10), as is usually the case of figures dancing in Minoan

Figure 3. Gold ring from Isopata, Crete (after Niemeier 1990, fig. 2).

art, for example the "Dancing Girl" fresco from the Queen's megaron at Knossos (Evans 1930, 70; Kontorli-Papadopoulou 1996, 43, 152). Naturally, the scene does not represent an orgiastic dance as Evans initially argued, but there is a sense of rhythm in the movement of the women (Immerwahr 1990, 66). It should be borne in mind that dance can also refer to the performance of gestures, as is the case in some cultures, for example the Chinese sleeve gestures or the Indian mudras (Royce Peterson 1977, 195), and therefore the possibility that this was also the case in the Prehistoric Aegean should not be rejected. Nanno Marinatos, for example, prefers to characterize the performance of the figures in the aforementioned fresco as "sacred mime", according to Groenewegen-Frankfort's expression (Marinatos 1987a, 141), which can be part of a dance (Royce Peterson 1977, 193-194).The female dancers/performers in the "Sacred Grove and Dance" fresco face to the left (with the exception of two figures that face to the right), as does the crowd, suggesting that they are watching something taking place to the left, rather than the dancers themselves (Evans 1930, 65; Davis 1987, 158; German 2005, 27). Therefore, in this case dancing is not performed for its own sake but as part of a ritual in which the aim is something else, probably the appearance of the divinity. Nanno Marinatos has argued that there is a terminal point towards which the women "dancing" are heading and this is probably some architectural feature, either an altar or a shrine (Marinatos 1987a, 141; 1987b, 24; Davis 1987, 158), which would be in accordance with the expectation of the epiphany of the divinity.

However, it is possible that the "Sacred Grove and Dance" fresco depicts a rite of passage and specifically the transition from one age group to another, rather than the epiphany of the divinity (Jacobs 2004). The "Sacred Grove and Dance" fresco and another miniature fresco known as the "Grandstand" or "Temple" fresco may have formed parts of one pictorial program, since they were uncovered in the same room (Evans 1930, 74; Cameron

Figure 4. "The Sacred Grove and Dance" fresco, reconstructed by N. Marinatos (after Marinatos 1987a, pl. 142, fig. 7).

1987, 325; Jacobs 2004, 15; German 2005, 27). Ariane Jacobs has interpreted the crowds in both scenes as representing different groups of people, indicated by their separation into male and female sections, by their dresses, hairstyles and jewelry (Jacobs 2004, 15). The relation of age and social status with hairstyles has been studied by Ellen Davis (1986) and Robert Koehl (1986). Therefore, it is possible that the standing female figures or at least some of them on the "Sacred Grove and Dance" fresco represent young girls, indicated by their shaved heads (Marinatos 1987b, 25; Jacobs 2004, 16). The fresco, then, would depict a ceremony celebrating the transition of children into youths (Jacobs 2004, 16-17). According to Jacobs, the actual ordeals performed in the rite of passage, for example bull-leaping, may have been depicted on a third frieze and the particular fresco probably represents only the celebration of their transition (Jacobs 2004, 16-17). However, the connection of dancing with rites of passage would not be unusual, since dance metaphorically can enact and communicate status transformation, whereas anthropologically it is often attested as part of initiation rites (Hanna 1987, 112).

The association of the fresco with rites of passage is perhaps further supported by the representation on the right of the composition of a group of male figures holding javelins faced by another male figure, an officer according to Davis (1987, 159). It is perhaps plausible to argue that both the female figures on the foreground and the male figures at the side represent groups of initiates set aside for that reason.

These dances, presumably as part of festivals, may have been performed in the west courts of the Minoan palaces, indicated by the depiction of the causeways, an actual feature of these courts of the palaces at Knossos, Phaistos and Mallia (Immerwahr 1983, 145; Davis 1987, 157; Marinatos 1987a, 141; Jacobs 2004, 11; Liveri 2008, 24). In the case of Phaistos there are also steps for standing observers on two sides of the court, whereas at Knossos such steps exist in the northern extension of the west court, known as the Theatral Area (Preziosi and Hitchcock 1999, 100; Jacobs 2004, 14; German 2005, 50).

The Harvesters' Vase (LM I) from Ayia Triadha on Crete could also refer to initiation rites (**fig. 5**). The scene decorates a stone rhyton –a ritual vessel- and depicts a procession composed mostly of young men with an older leader carrying hoes and chanting (Marinatos 1987b, 27; Younger 1988, 2, 6). The assumption that they may be performing some kind of dance is based on the fact that they have one leg up, indicating a vigorous rhythm, whereas the depiction of a seistron and of four figures with open mouths indicates the presence of music and singing (Younger 1988, 8). A clay model representing a circular dance dating to Middle Minoan III (*c.* 1650 BC) from the tholos tomb at Kamilari could also refer to initiation rites. The model depicts four male figures dancing in a close circle with arms touching. The figures are naked, except from a conical hat they are wearing, and they are depicted dancing in a closed circular area decorated with horns of

Figure 5. The Harvesters' Vase from Ayia Triadha, Crete (after the Archaeological Museum of Herakleion guide, p. 130).

consecration (Mandalaki 2003, 29; 2004, 16; Liveri 2008, 21), a religious symbol (Nilsson 1950, 165).

The performance of dance may also be depicted on other Minoan frescoes. A number of women depicted on the miniature frescoes from Tylissos (LM I) may be interpreted as dancing, on the basis of their festal garments and gestures, similar to those on the "Sacred Grove and Dance" fresco, or may simply be making preparations for a feast (Immerwahr 1983, 146; Kontorli-Papadopoulou 1996, 52, 153). Fragments of a miniature fresco uncovered in area M at Ayia Irini on the island of Keos (LM IB, *c.* 1500-1450 BC) show the composition of a hillside town and its inhabitants (Abramovitz 1980, 57). At least 19 men are taking part in a procession, dressed in long white robes or hide/wool skirts and carrying offerings. In four other fragments men and women seem to be dancing (Peterson 1981, 87; Morgan 1988, 94, 98, fig. 61; 1994, 243; Kontorli-Papadopoulou 1996, 59). The figures were interpreted as dancing in a row, perhaps participating in a festival (Abramovitz 1980, 58; Morgan 1988, 98). The mouths of some men are open, as if singing, and have both arms raised with open hands, gesturing (Abramovitz 1980, 58; Kontorli-Papadopoulou 1996, 59, 153). These frescoes could suggest that dancing was not only associated with epiphany or initiation rituals, but might have been performed in other occasions as well, often with the accompaniment of singing.

Finally, it is worth mentioning that particular spaces and structures have been associated with dancing as early as the Early Minoan period (*c.* 3100-2000 BC). Specifically, open spaces associated with Early Minoan tholos tombs have been interpreted as places where dancing could have taken place (Branigan 1970, 135; 1998, 21). Such areas, which were specially surfaced, were found east of the doorway of Tomb E and in front of Tomb B at Koumasa, between the two large tholoi at Platanos and in front of the three built chamber tombs at Mochlos (Branigan 1970, 135; Xanthoudides 1971, 9, 90). According to Branigan none of the funerary rites (libations, deposition of offerings) could justify the construction of these large paved areas outside the tombs. Therefore, he identified these areas as the precursors of the central and western courts of the Minoan palaces and hence as places for ritual dancing (Branigan 1970, 135; 1998, 21).

In terms of the performance of dancing in the Minoan palaces, in addition to west courts, dancing might have also taken place in actual dancing places. Three low circular structures were discovered in the palace of Knossos, uncovered some 350m west of the northern edge of the palace and dating to the early LM IIIA1 period (*c.* 1400-1350 BC). The structures, which have a smoothed paved surface, were interpreted as dancing places (Warren 1984; 1988, 9), associated most likely with ceremonies taking place in the palace. It is true that no archaeological evidence which might indicate the function of these structures was discovered and their association with dance was based on iconographic parallels of circular dance, such as the Palaikastro model and the existence of historical evidence associating Crete with a very old dancing tradition (German 2005, 50). However, the interpretation is plausible.

Turning to Mycenaean Greece, depictions of dancing are scarcer and seem to come mostly from funerary contexts. The Tanagra larnakes dating to LH IIIA-C (*c.* 1400-1100 BC) provide interesting information concerning the performance of funerary/post funerary rites, of which dance was part. The choice of subject matter and the style of drawing of the Tanagra larnakes are very different from the contemporary Cretan examples, indicating that, although Crete was the source for the idea and the forms of larnakes, the choice of subject matter and style of drawing were Mycenaean (Vermeule 1965, 136-137; Immerwahr 1995, 109). Specifically, a circular dance is depicted on the long side of the Tanagra larnax from Tomb 22 (**fig. 6**), suggested by the alternation of female figures in red and black (Spyropoulos 1974, 24-25; Gallou 2005, 37). The figures are dressed in a baggy plain dress and have both hands raised to their heads, a gesture clearly expressing grief.

Dancing seems also to be depicted on the larnax from Tomb 10 (LM IIIA1, *c.* 1390-1370 BC) from the cemetery of Mochlos on Crete. Two figures are painted just beneath the rim with their hands on their waist, suggesting that they may be dancing, each wearing pointed shoes and possibly a dagger and one of them wearing an animal mask, representing a jackal (Brysbaert 2001).

A fresco from Ayia Triadha (LM IIIA1) is so far the only probable Mycenaean representation of dance in large scale art. The lower register of the fresco depicts six or seven women with their arms extended and touching the shoulders of the figure in front, who have been interpreted as engaged in a ritual dance (Immerwahr 1990, 102; Kontorli-Papadopoulou 1996, 51, 152).

The iconography of gold rings, often related to ecstatic dances, has already been examined in relation to Minoan ritual dancing. Although the iconography is clearly Minoan, it seems that the Mycenaeans either imported gold rings with specific scenes and/or adopted the gold rings produced on the mainland with the elements which had meaning for them in terms of their religion (Niemeier 1990, 166, 170). For example, the scene depicted on the gold ring from Chamber tomb 91 at Mycenae represents the performance of dancing, but the individuals involved are all human beings. The central female figure has both arms to the waist in a similar manner to that of the statues from the temple at Ayia Irini, whereas there are no floating objects next to her. According to Niemeier, then, she is a human being, rather than a goddess (Niemeier 1990, 169). Therefore, in this case a Minoan motif of ecstatic epiphany is transformed in Mycenaean iconography as a ritual where only human beings are involved (Niemeier 1990, 169; Palikisianos 1996, 834).

So far the only Mycenaean clay model representing dancers is found in the British Museum collection (Hirsch and Virsi Collections, catalogue number GR1996.3-25.1). It consists of a terracotta group of three figures dancing in a circle with arms stretched. The object is dated to LH IIIA2-B1 (*c.* 1390-1200 BC) and the figures have the typically Mycenaean bird-like head and stripes decorating their dress. Unfortunately, the context in which this model was found is unknown.

A male figure dressed in hide skirt and interpreted as engaged in wild dancing (**fig. 7**) is depicted on a krater (LH IIIC) from chamber tomb 5 from the Ayia Triadha cemetery at Elis (Schoinas 1999, 258). That the figure seems to be engaged in a wild dance is suggested by the raised arms, but mostly by the fact that his feet are raised off the ground (Schoinas 1999, 259). According to the excavator, the figure may be dancing a sacred dance before the sacrifice of the animal depicted behind the two female figures, who are probably also priestesses, whereas the fact that he is holding a stone axe further supports his involvement with animal sacrifice (Schoinas 1999, 258-259). However, the dance

Figure 6. Larnax with the representation of a dance from Tomb 22 at Tanagra (after Immerwahr 1990, pl. XXI).

performed by the male figure may in general be related to the funerary rites carried out in honour of the deceased.

A dance is also painted on an hydria dating to LH IIIC (*c.* 1190-1065 BC) found in the cemetery of Kamini at Naxos representing eight figures dancing in a circular formation, some of them also represented as having an animal head (**fig. 8**). Specifically the fourth and fifth figure seem to have the head of a horse, the sixth that of a bird and the eighth that of a cat (Mastrapas 1996, 798-799). The excavator has interpreted the vessel as a ritual object for storing a special kind of drink used in libation in honour of the dead (Mastrapas 1996, 798, 799). Depictions of figures wearing masks are absent from Mycenaean Greece and Minoan Crete, with the exception of the aforementioned hydria and a seal also from Naxos depicting a figure probably wearing a bird mask (Kardara 1977, 6). The absence of other representations does not allow us to reach any conclusions as to the occasion or the identity of the persons, but it is possible that both the figures on the Naxos hydria and the seal were members of the priesthood. It is very interesting that figures in masks are represented in association with a funerary context. Perhaps the scene represents the narration of a myth, or the aim of the dance was apotropaic. Whichever their meaning, the use of masks in dances can be understood as part of the transmission of meaning -a feature also of dances- through the transformation of the wearer (Tonkin 1992, 225).

Dance and the Prehistoric Aegean

Dance in the prehistoric Aegean was closely associated with the performance of cult. Dancing was considered as an appropriate way to invoke and celebrate the appearance of the divinity, but was also performed in initiation rites, in rites connected to the dead, and perhaps even in secular celebrations. Nevertheless, dance, and performance in general, could be used as a way of stating identities,

Figure 7. Prothesis scene on a LH IIIC krater from Chamber Tomb 5 from the cemetery of Ayia Triadha, Elis (after Schoinas 1999, p. 798, fig. 1).

Figure 8. Dance scene on a LH IIIC hydria from Kamini, Naxos (after Mastrapas 1996, p. 798, fig. 1 and 2).

establishing and maintaining authority. An example of how dance may have been used to state identity and legitimize land ownership can be provided by the association of Early Minoan tholoi tombs in the plain of Mesara with the dance places mentioned above. During the Early Minoan period, elaborate funerary and/or post-funerary rites were performed in tombs and specifically tholoi in order to legitimise and emphasise the authority of particular groups over the land (Murphy 1998). The choices made by a society concerning the disposal of their dead reflect, in addition to the beliefs, the ideologies and the social structure of the particular society (Murphy 1998, 27). The more ritualised the disposal, the greater its significance to the living members of the society. Based on that notion it has been argued that Pre-palatial Minoan tholoi tombs functioned as territorial markers that served as claims by the living to linear descent from the deceased buried in the tombs, as well as to establish the rights of the residents in settlements associated with these tombs to resources in their vicinity (Murphy 1998). In order to achieve this, all three stages associated with death (separation, transition, but mostly incorporation) and hence the transformation from individual to ancestor, were accompanied by rituals (Murphy 1998, 32-34). In this context, the performance of dancing could serve two purposes: first of all to celebrate the integration of the deceased into the realm of the ancestors in a very straight-forward way, and secondly to reunite lineage ties that had been disrupted by death, as well as to enhance the solidarity of the particular community and more importantly to exclude people who were not members of it. Moreover, it should be noted that Branigan has suggested that ceremonies performed in Early Minoan tholos cemeteries on the plain of Mesara were not restricted to funerary rites, since cemeteries served as foci of ritual performances in general due to their role as markers of the community's stability (Branigan 1998, 21).

With the construction of the first palaces and the establishment of a firm hierarchy, there was a shift in the location where dance was performed, from the tombs to the palaces (Branigan, in German 1999, 281). Senta German has interpreted dance as part of social dramas, that is "a series of structured social events which address crisis in a culture"[3] functioning as a means of social control

---

[3] For a detailed examination of the concept of "social dramas", see Turner 1974.

during times of political change and upheaval (German 2005, 85). In the Neopalatial period such social dramas helped the establishment and maintenance of new social groupings. In the Pre-palatial and Proto-palatial period kinship relations unified people, whereas in the Neopalatial period extra-familial social authority had central control (German 2005, 92). This centralisation of control during the Neopalatial period is also suggested by the centralisation of cult practices, with the reduction in the number of peak sanctuaries and cult sites around the palaces (Peatfield 1987; Gesell 1987). Cult during the Neopalatial period was brought inside the palaces and its access was controlled (Gesell 1987, 126; German 2005, 92).

Evidence for dance in Mycenaean Greece is not as abundant as for Minoan Crete and the fact that the Great Courts of the mainland palaces were designed to control access and accommodate a limited number of people (Cavanagh 2001, 130) suggests the performance of festivals of a different character to Minoan ones. So far, there is no clear evidence to indicate that dancing in Mycenaean Greece was part of public festivals, which were so important to the Minoans. However, depictions of priests dancing, such as the male figure on the krater from the cemetery at Elis and most likely the central female figure on the gold ring from Mycenae (Vasilikou 1997, 56) indicate that dancing was part of Mycenaean cult.

In terms of the performance of dancing in relation to death, Yannis Hamilakis has interpreted this as part of a whole mechanism aimed to generate remembering and forgetting, such as food and drink consumption, secondary burial treatment and the 'killing' of memory by the 'killing' of artefacts. All these actions were parts of a multifaceted social phenomenon implicated in the construction of social persons, which evoked senses, emotions and feelings (Hamilakis 1998, 128). Performative ceremonies generate bodily sensory and emotional experiences, which result in habitual memory becoming sedimented in the body (Hamilakis 1998, 117). Food and drink consumption is the most obvious practice, but under this light it becomes clear why dance may have formed part of such rituals.

Conclusion

Dance would have been chosen as part of performance due to its capacity to cause a response on the part of the audience, which in some cases could have been intentionally provoked; to convey meaning at different levels, perceived by all five senses and to conceal true intentions. Dance in the Prehistoric Aegean may have been a part of religious ceremonies for the veneration of the divine, of rites of passage marking the transition from one age group to another, or in death rites to celebrate the incorporation of the dead into the sphere of the ancestors. At the same time, dance may have been used in order to communicate complex messages concerning social circumstances and relations, as well as relations of power and subordination.

The aim of this paper was clearly not to exhaust the topic but to contribute to the study of dance in the Prehistoric Aegean by emphasising the significance of dance not only as part of religious ceremonies, but also as a way of expressing social and political realities and as a potential means of political control.

References

Aamodt, C. 2006. *Priests and Priestesses in Mycenaean Greece*. Unpublished PhD thesis, University of Nottingham.

Abramovitz, K. 1980. Frescoes from Ayia Irini, Keos. Parts II-IV. *Hesperia* 49, 57-85.

Blacking, J. 1985. Movement, dance, music, and the Venda girls' initiation cycle. In P. Spencer (ed.), *Society and the dance: The social anthropology of process and performance*, 64-91. Cambridge, University Press.

Branigan, K. 1970. *The Tombs of Mesara*. London, Duckworth.

Branigan, K. 1993. *Dancing with Death: life and death in Southern Crete, c. 3000-2000 BC*. Amsterdam, Adolf M. Hakkert.

Branigan, K. 1998. The Nearness of You: Proximity and Distance in early Minoan Funerary Landscapes. In K. Branigan (ed.), *Cemetery and Society in the Aegean Bronze Age*, 13-26. Sheffield, Studies in Aegean Archaeology.

Brinson, P. 1985. Epilogue: Anthropology and the study of dance. In P. Spencer (ed.), *Society and the Dance. The social anthropology of process and performance*, 206-214. Cambridge, University Press.

Brysbaert, A. 2001. A Bronze Age larnax from Crete Revived. Where Old and New meet on Crete. *Anistoriton* 2001. http://www.anistor.gr/english/enback/p001.htm

Burkert, W. 1985. *Greek Religion*. Oxford, Blackwell Publishers.

Cameron, M. A. S. 1987. The 'Palatial' Thematic System in the Knossos Murals, in R. Hägg and N. Marinatos (eds.), *The Function of the Minoan Palaces. Proceedings of the Fourth International Symposium at the Swedish Institute in Athens, 10-16 June, 1984*, 321-325. Stockholm, Paul Åströms Förlag.

Caskey, M. E. 1986. *KEOS II. The Temple at Ayia Irini. Part I: The Statues*. Princeton, American School of Classical Studies.

Cavanagh, W. C. 2001. Empty Space? Courts and Squares in Mycenaean Towns. In K. Branigan (ed.), *Urbanism in the Aegean Bronze Age*, 119-134. Sheffield, Studies in Aegean Archaeology.

Davis, E. N. 1986. Youth and Age in the Thera Frescoes. *American Journal of Archaeology* 90, 399-406.

Davis, E. N. 1987. The Knossos Miniature Frescoes and the Functions of the Central Courts, in R. Hägg and N. Marinatos (eds.), *The Function of the Minoan Palaces. Proceedings of the Fourth International Symposium at the Swedish Institute in Athens, 10-16 June, 1984*, 157-161. Stockholm, Paul Åströms Förlag.

Dragona-Latsoudi, A. 1977. Μηκυναϊκός Κιθαρωδόςαπότην Ναυπλία. *Archaeologiki Efimeris* 1977, 86-98.

Doumas, C. 1965. Κορφή Τ'Αρωνιού. *Archaeologikon Deltion* 20, 41-64.

Evans, A. 1930. *The Palace of Minos Vol. III.* London, MacMillan and Co Ltd.

Gallou, C. 2005.*The Mycenaean Cult of the Dead.* British Archaeological Reports International Series 1372. Oxford, BAR Publishing.

German, S. C. 1999. The politics of dancing: a reconsideration of the motif of dancing in Bronze Age Greece, in P. Betancourt, V. Karageorghis, R. Laffineur and W-D.Niemeier (eds.), *Meletemata. Studies in Aegean Archaeology Presented to Malcolm H. Wiener as he enters his 65th year. Vol.I.* Aegaeum 20, 279-281. Liége, Annales d'archéologie égéenne de l'Université de Liége et UT-PASP.

German, S. C. 2005. *Performance, Power and the Art of the Aegean Bronze Age.* British Archaeological Reports Int. Series 1347. Oxford, BAR Publishing.

Gesell, G. C. 1987. The Minoan Palace and Public Cult, in R. Hägg and N. Marinatos (eds.), *The Function of the Minoan Palaces. Proceedings of the Fourth International Symposium at the Swedish Institute in Athens, 10-16 June, 1984*, 123-128. Stockholm, Paul Åströms Förlag.

Goodison, L. and Morris, C. 1998. Beyond the 'Great Mother': The Sacred World of the Minoans, in L. Goodison and C. Morris (eds.), *Ancient Goddesses*, 113-132. London, British Museum Press.

Hägg, R. 1983. Epiphany in Minoan Ritual. *Bulletin of the Institute of Classical Studies* 30, 184-185.

Hägg, R. 1986. Die göttliche Epiphanie im minoischen Ritual. *Mitteilungen des Deutchen Archäologischen Instituts, Athenische Abteilung* 101, 46-62.

Hamilakis, Y. 1998. Eating the Dead: Mortuary Feasting and the Politics of Memory in the Aegean Bronze Age Societies. In K. Branigan (ed.), *Cemetery and Society in the Aegean Bronze Age*, 115-132. Sheffield, Studies in Aegean Archaeology.

Hanna, J. L. 1987. *To dance is Human. A Theory of Nonverbal Communication.* Chicago, University Press.

Immerwahr, S. A. 1983. The people in the frescoes, in O. Krzyszkowska and L. Nixon (eds.), *Minoan Society. Proceedings of the Cambridge Colloquium 1981*, 143-153. Bristol, Classical Press.

Immerwahr, S. A. 1990. *Aegean Painting in the Bronze Age.* Philadelphia, Pennsylvania State University Press.

Immerwahr, S. A. 1995. Death and the Tanagra larnakes, in J. B. Carter and S. P. Morris (eds.), *The Ages of Homer*, 109-121. Austin, University of Texas Press.

Jacobs, A. 2004. The Knossos Miniature Paintings Reconsidered. *Journal of Prehistoric Religion* XVIII, 8-20.

Kardara, C. 1977. Απλώματα Νάξου. Κινητά ευρήματα Τάφων Α και Β. Βιβλιοθήκη της εν Αθήναις Αρχαιολογικής Εταιρείας, Αρ. 88.

Koehl, R. B. 1986. The Chieftain Cup and a Minoan Rite of Passage. *Journal of Hellenic Studies* CVI, 99-110.

Kontorli-Papadopoulou, L. 1996. *Aegean Frescoes of Religious Character.* SIMA CXVII. Göteborg, Paul Åströms Förlag.

Lawler, L. B. 1965. *The Dance in Ancient Greece.* Wesleyan, University Press.

Liveri, A. 2008. Representations and Interpretations of Dance in the Aegean Bronze Age. *Mitteilungen des Deutchen Archäologischen Instituts, Athenische Abteilung* 123, 1-38.

Lonsdale, S. H. 1993. *Dance and Ritual Play in Greek Religion.* Baltimore, The John Hopkins University Press.

Mandalaki, S. 2003. Ο χορός στον Προϊστορικό Ελλαδικό Χώρο, in E. Andrikou, A. Goulaki-Voutyra, C. Lanara and Z.Papadopoulou (eds.), *Mouson Dora: Mousikoi kai choreutikoi apoichoi apo tin archaia Ellada. Gifts of the Muses: Echoes of Music and Dance from Ancient Greece*, 27-31.Athina, Ypourgeio Politismou.

Mandalaki, S. 2004. Ο χορός στη Μινωϊκή Κρήτη. *Arhaiologiakai Tehnes* 90, 15-20.

Maners, L. D. 2006. Utopia, Eutopia and EU-topia: Performance and Memory in Former Yugoslavia. In T. J. Buckland (ed.), *Dancing from Past to Present: Nation, Culture, Identities*, 75-96. Wisconsin, University Press.

Marinatos, N. 1987a. Public Festivals in the West Courts of the Palaces, in R. Hägg and N. Marinatos (eds.), *The Function of the Minoan Palaces. Proceedings of the Fourth International Symposium at the Swedish Institute in Athens, 10-16 June, 1984*, 135-143. Stockholm, Paul Åströms Förlag.

Marinatos, N. 1987b. Role and Sex Division in Ritual Scenes of Aegean Art. *Journal of Prehistoric Religion* I, 23-34.

Marinatos, N. 1993. *Minoan Religion: Ritual, Image and Symbol.* Columbia, South Carolina University Press.

Mastrapas, A. N. 1996. Υδρία με ηθμωτό κυάθιο από το ΥΚ/ΥΕΙΙΙΓ νεκροταφείο Καμινιού Νάξου, in E. DeMiro, L. Godart and A. Sacconi (eds.), *Atti e Memorie del Secondo Congresso Internazionale di Micenologia, Roma-Napoli 14-20 ottobre 1991, Vol. 2: Storia. Incunabula Graeca* 98, 797-803. Rome, Gruppo editorial internazionale.

McFee, G. 1992. *Understanding Dance.* New York, Routledge.

Morgan, L. 1988. *The Miniature Wall Paintings of Thera.* Cambridge, University Press.

Morgan, L. 1994. The Wall-Paintings of Ayia Irini, Kea. *Bulletin of the Institute of Classical Studies* 40, 243-244.

Mouratidis, J. 1989. Are There Minoan Influences on Mycenaean Sports, Games and Dances? *Nikephoros* 2, 43-63.

Murphy, J. M. 1998. Ideologies, Rites and Rituals: A View of Prepalatial Minoan Tholoi. In K. Branigan (ed.), *Cemetery and Society in the Aegean Bronze Age*, 27-40. Sheffield, Studies in Aegean Archaeology.

Niemeier, W-D. 1987. On the Function of the "Throne Room" in the Palace at Knossos, in R. Hägg and N. Marinatos (eds.),*The Function of the Minoan Palaces. Proceedings of the Fourth International Symposium at the Swedish Institute in Athens, 10-16 June, 1984*, 163-168.Stockholm, Paul Åströms Förlag.

Niemeier, W-D. 1990. Cult Scenes on Gold Rings from the Argolid, in R. Hägg and G. C. Nordquist (eds.),*Celebrations of Death and Divinity in the Bronze Age Argolid. Proceedings of the Sixth International Symposium at the Swedish Institute at Athens, 11-13 June, 1988*, 165-170. Stockholm, Paul Åströms Förlag.

Nilsson, M. P. 1950. *The Minoan-Mycenaean Religion and Its Survival in Greek Religion*. New York, Biblo and Tannen.

Palikisianos, M. E. 1996. Θρησκευτικές Παραστάσεις στις Κρητομυκηναϊκές Σφραγίδες και Σφραγιστικά Δαχτυλίδια, in E. deMiro, L. Godart and A. Sacconi (eds.), *Atti e Memorie del Secondo Congresso Internazionale di Micenologia, Roma-Napoli 14-20 ottobre 1991, Vol. 2: Storia. Incunabula Graeca 98*, 833-844.Rome, Gruppo editorial internazionale.

Peatfield, A. 1987.Palace and Peak: The Political and Religious Relationship between Palaces and Peak Sanctuaries, in R. Hägg and N. Marinatos (eds.), *The Function of the Minoan Palaces. Proceedings of the Fourth International Symposium at the Swedish Institute in Athens, 10-16 June, 1984*, 89-93. Stockholm, Paul Åströms Förlag.

Peterson, S. E. 1981. *Wall Paintings in the Aegean Bronze Age: the Procession Frescoes*. Unpublished thesis, University in Minnesota.

Preziosi, D. and L. A. Hitchcock, 1999.*Aegean Art and Architecture*. Oxford, University Press.

Renfrew, C. 1985. *The Archaeology of Cult. The sanctuary of Phylakopi*. British School at Athens, Suppl. 18. London, Thames and Hudson.

Ρεθεμιωτάκης, Γ. 1998. Ανθρωπομορφική πηλοπλαστική στην Κρήτη. Από την νεοανακτορική έως την υπομινωική περίοδο.Athens 1998.

Royce Peterson, A. 1977. *The Anthropology of Dance*. Indiana, University Press.

Sakellarakis, J. A. 1967. Minoan Cemeteries at Arkhanes. *Archaeology* 20, 276-281.

Sakellariou, A. 1964. Die minoischen und mykenischen Siegel des Nationalmuseums in Athen. Corpus der minoischen und mykenischen siegel I. Berlin, Gebr. Mann Verlag.

Schiering, W. 1999.Goddess, Dancing and Flower-Gathering Maidens in Middle Minoan Vase Painting, in P. Betancourt, V. Karageorghis, R. Laffineur and W-D.Niemeier (eds.), *Meletemata. Studies in Aegean Archaeology Presented to Malcolm H. Wiener as he enters his 65th year. Vol. III* (Aegaeum 20), 747-749. Liége, Annales d'archéologie égéenne de l'Université de Liége et UT-PASP.

Schoinas, C. 1999. Εικονιστική παράσταση σε όστρακα κρατήρα από την Αγία Τριάδα Ηλείας. In Η Περιφέρεια του Μυκηναϊκού Κόσμου, Α' Διεθνές Διεπιστημονικό Συμπόσιο, Λαμία 25-29 Σεπτεμβρίου 1994, 257-262. Υπουργείο Πολιτισμού & ΙΔ' ΕΠΚΑ.

Sklar, D. 2006. Qualities of Memory: Two Dances of the Tortuga Fiesta, New Mexico. In T. J. Buckland (ed.), *Dancing from Past to Present: Nation, Culture, Identities*, 97-122.Wisconsin, University Press.

Soar, K. 2010. Circular Dance Performances in the Prehistoric Aegean, in A. Michaels *et al.* (eds.), *Ritual Dynamics and the Science of Ritual*, 137-156. Wiesbaden, Harrasowitz Verlag.

Spencer, P. 1985. Introduction. In P. Spencer (ed.), *Society and the dance: The social anthropology of process and performance*, 1-46. Cambridge, University Press.

Spyropoulos, T. G. 1974. Ανασκαφή Μυκηναϊκής Τανάγρας. Πρακτικά της εν Αθήναις Αρχαιολογικής Εταιρείας 1974, 9-33.

Tonkin, E. 1992. Mask. In R. Bauman (ed.), *Folklore, Cultural Performances and Popular Entertainment: A Communications-Centred Handbook*, 225-232. Oxford, University Press.

Turner, V. 1974. *Dramas, Fields and Metaphors: Symbolic Action in Human Society*. Cornell, University Press.

Vasilakis, A. 2000. Archaeological Museum of Herakleion. Athens, Adam editions.

Vasilikou, N. 1997. Μυκηναϊκά Σφραγιστικά Δαχτυλίδια. Βιβλιοθήκη της εν Αθήναις Αρχαιολογικής Εταιρείας αρ. 166. Athens.

Vermeule, E. T. 1965. Painted Mycenaean Larnakes. *Journal of Hellenic Studies* LXXXV, 123-148.

Warren, P. 1984. Circular platforms at Minoan Knossos. *British School at Athens* 79, 307-323.

Warren, P. 1988. Minoan Religion as Ritual Action. Gothenburg, University Press.

Williams, D. 2004. *Anthropology and the Dance*. Illinois, University Press.

Xanthoudides, S. 1971. *The Vaulted Tombs of Mesara*. England, Gregg International Publishers Ltd.

Younger, J. G. 1988. *Music in the Aegean Bronze Age*. Sweden, Paul Åströms Förlag.

THE DANCE OF DEATH:
DANCING IN ATHENIAN FUNERARY RITUALS

Hugh Thomas

University of Sydney

Abstract: A small passage sung by the chorus in Euripides, Heracles mentions a practice known as 'the dance of death'. This reference provides the only literary evidence of the role of dancing within a funerary context in ancient Greece. This article seeks to address and discuss the evidence for the existence of dancing in Athenian funerary customs. It will focus on an analysis of funerary iconography found on Geometric, Protoattic and Black Figure pottery, which provides valuable insight in hitherto unknown Athenian funerary practices.

> "Alas! What groans or wails, what funeral dirge, or dance of death am I to raise?"
>
> Euripides, *Heracles*, 1025-1027

Euripides tragic play, *Heracles*, dramatically recounts the murder of Heracles' family by the hero, during a state of divine imbued frenzy. The chorus, lamenting the deaths of the hero's family, offer three different forms of mourning: wailing, singing a dirge, and dancing. Both wailing and singing are known to have been common forms of mourning during Greek funerary rituals. In the *Iliad*, both practices are attested to during the death of Hector.

"They laid him out on a ... bed, and they set beside him singers, leaders of the *threnoi*, who led the song of lamentation, they sang the dirge, and at this the women wailed. And among them white-armed Andromache led the lament, holding the head of man-slaying Hektor in her arms" (Homer, *Iliad*, 24.719-24).

Likewise, wailing and funerary dirges are commonly depicted, or at least alluded to, in Athenian funerary iconography. Arguably the most emotive representation of this comes from a Black Figure funerary plaque from the Sappho Painter dating to c.500 BC (Paris, Musee du Louvre: MNB905). The plaque is illustrated with mourning family members, many of whose relationship to the deceased is inscribed next to them. Further inscriptions of 'Oimoi', which translates to 'Alas', are found between the figures, thus documenting their cries of sorrow.

Yet the practice of dancing as a funerary ritual is rarely alluded to in ancient literature and when it is, it is often due to unusual circumstances. Plutarch records that after the death of Aratus of Sicyon, his troops consulted the Delphic oracle who commanded them to bury Aratus inside their city walls. In response to this, "the Sicyonians, in particular, changing their mourning into festival, at once put on garlands and white raiment and brought the body of Aratus from Aegium into their city, amid hymns of praise and choral dances; and choosing out a commanding place, they buried him there, calling him founder and saviour of the city" (Plutarch, *Aratus* **53.1-5**). In this instance, the traditional mourning practices of the Sicyonians were abandoned in favour of celebration, and as a result, dancing is not part of a funerary ritual.

So the question remains as to whether dancing occurred during Athenian funerary rituals. The 'dance of death' alluded to by the chorus in *Heracles* could be a result of the chorus' primary role within theatrical productions, with dancing a major aspect of this. In fact, the location of the chorus within the theatre was limited to the *Orchestra*, which in itself means 'dancing place' (Weiner 1980, 205).

Yet despite the relative lack of literary evidence pertaining to the role of dancing at funerals, two groups of Athenian funerary pots appear to show figures dancing in a funerary context. The first group comprises of a series of Late Geometric (760-700 BC) and Protoattic (c.700-630 BC) period Hydriai decorated with dancing women, possibly taking part in a post-burial ritual. The second consists of two Black Figure Kantharoi illustrated with dancing hoplites taking part in an *ekphora*, the transportation of the deceased to the cemetery. Several other non-Attic vessels also appear to show dancing in a similar fashion, yet this paper will focus primarily on Athenian examples due to the vast wealth of comparable funerary imagery and literary sources.

Dancing on Geometric and Protoattic Pottery

In c.900 BC, Athens entered the 'Geometric Period', referenced to as such because of the prominence of Geometric motifs in Athenian art. Pottery was decorated with small bands of Geometric patterns, which over the ensuing century spread to cover the entire vessel.[1] However, it was not until the end of Middle Geometric II (800-760 BC) that Athenian potters began to dramatically develop the painting of the human figure. Land and sea battles became a prominent theme, as did funerary iconography.

The vast majority of Geometric funerary pottery was illustrated with *prothesis* scenes, the ritualised laying out of the deceased. Plato discusses the original purpose of the *prothesis*, stating, "as to the laying-out of the corpse, first, it shall remain in the house only for such a time as is required to prove that the man is not merely in a faint, but really dead" (Plato, *Laws* 959a). It is known that this event took place on the day after the death, the second day, as documented by Antiphon, who wrote, "On the first day, the day of the boy's death, and on the second, when the body was laid out" (Antiphon, *On the Choreutes* 6.34). We know that in Archaic and Classical Greek funerary rituals, men and women had distinct roles during the *prothesis*. Women were the primary carers for the deceased and as such sacrificed themselves to the pollution that was coupled with such rituals (Demosthenes, *Against Macartatus* 62-65).[2] In contrast, men played a more ancillary role and took no part in the preparation or the primary mourning of the deceased.

The next stage in the process was the *ekphora*, the transportation of the deceased to the grave (Ahlberg 1971, 220-239; Shapiro 1991, 630-634). In the Archaic and Classical periods the *ekphora* occurred before dawn on the third day after death. Solonian reforms of the early 6th century BC dictated that the *ekphora* was to take place before sunrise and that the procession must travel through side streets and remain relatively quiet (Plutarch, *Solon* 21). This implies that prior to these reforms the *ekphora* had been a more public, grandiose and an altogether noisier affair. The *ekphora* wound its way to the cemetery, followed by a procession of mourning relatives. During the 6th and 5th centuries BC, mourners could be hired for the event to help increase the prestige of the deceased. Plato mentions these mourners, stating "just as a corpse is escorted with Carian music by hired mourners" (Plato, *Laws* 800). Imagery of the *ekphora* is relatively rare in Athenian art, although several Late Geometric vases decorated with the motif have been discovered.

The importance of both the *prothesis* and *ekphora* for the Athenians is clear, as a number of monumental vessels decorated with imagery of the two rituals were produced during the Late Geometric Period. These vases were often populated with dozens of mourning figures, bands of chariots, warriors, and representations of the deceased. However, towards the end of the 8th century another possible funerary scene began to appear. The motif consisted of women, and the occasional man, standing in long lines whilst holding hands. This subject continues into the Protoattic period (7th century BC) but disappears by the middle of the century. Despite the lack of movement prevalent in later dancing iconography, it is clear that figures who are shown holding hands are dancing, as several of the vessels (**figs. 1 and 2**) are illustrated with dancers following musicians (Langdon 2008, 158ff). Further evidence comes from figures depicted, as Langdon states, in "rhythmic compositions" or who are shown with "specific body positions...and particular details of dress and prop" (Langdon 2008, 158). Apart from clasping hands, the women are also found holding branches and wreaths, whilst others have rows of upturned chevrons descending from their hands.

The most significant issue with these vessels are whether they can be considered funerary in nature. The gesture of holding hands does not appear on any pot decorated with definitive funerary imagery, like a *prothesis* or *ekphora*. However, it is shown in conjunction with motifs such as files of chariots and warriors, which commonly decorate funerary vases (**fig. 3**). Furthermore, the vessels are also frequently adorned with plastic snakes, attached to the body, neck and handles of the vessel (fig. 1 &2). The serpent has a specific connection with the dead, as Plutarch asserts:

> "as putrefying oxen breed bees, and horses wasps, and as beetles are generated in asses which are in the like condition of decay, so human bodies, when the juices about the marrow collect together and coagulate, produce serpents. And it was because they observed this that the ancients associated the serpent more than any other animal with heroes" *(*Plutarch, *Cleomenes* 39, 3*)*

According to Garland, "On vases they, (snakes), are represented as inhabiting (or painted upon) graves, either symbolising the dead or else protecting them" (Garland 2001, 158-159). Furthermore, snakes are commonly found on Geometric vessels that are funerary in nature (fig. 3). The mortuary role of serpents is also found on a Black Figure Kantharoi discussed below, as it is illustrated with a grave clearly decorated with a snake (**fig. 4**).

Further evidence for the funerary nature of the gesture may come from one fragmentary Protoattic Hydria from the Athenian Agora (fig. 3), which appears to depict the figures dancing around a strange object. Papadopoulos argues this object is floral, although conversely it could also be interpreted as an amphora (Papadopolous 2007, 142). Amphorae were commonly used as grave markers during the Geometric period and it is possible that this is what the object represents.

---

[1] For the development of Geometric Pottery, see Coldstream 2003.
[2] For pollution in Greek funerary rituals, see Kurtz & Boardman 1971, 149; Parker 1983, 32-40; Morris 1987, 192-193; Garland 2001, 41-47.

Figure 1. Athenian Geometric Hydria with Plastic Snakes. Museum of Classical Archaeology, Cambridge inv.345 (Photo used with permission by Museum Of Classical Archaeology, Cambridge).

Figure 2. Protoattic hydria adorned with plastic snakes and illustrated with dancers and musicians. Agora P 10154. (Water colour image by Piet de Jong, photographed by C.Mauzy. Image used with permission by the American School of Classical Studies at Athens, Agora Excavations).

Figure 3. Athenian Geometric Amphora adorned with plastic snakes and illustrated with a prothesis. Agora P 4990. (Photographed by C.Mauzy. Image used with permission by the American School of Classical Studies at Athens, Agora Excavations).

To date, the most thorough research on the dancing figures is found in Susan Langdon's book, *Art and Identity in Dark Age Greece, 1100-700 B.C.E,* in which she analyses 50 Attic Late Geometric and Protoattic vases decorated with dancing figures (Langdon 2008, 168-74). Langdon suggests that the addition of plastic snakes may have had "more to do with water", due to their regular appearance on Hydriai, or perhaps they were only added to some vessels so as to be "appropriate to both funerary and non-funerary contexts" (Langdon 2008, 172). Indeed, on vessels not adorned with snakes, the gesture appears on eight different pot shapes,

Figure 4. Athenian Black Figure Kantharos from the Perizoma Group. Bibliothèque nationale de France Vase 353. (Photo provided with permission from the Bibliothèque nationale de France).

including, tankards, Hydriai and Skyphoi whilst those with snakes are almost all Hydriai (Langdon 2008, 169-170, Table 3.3). Langdon's suggestion that some vessels were made for funerary purposes and others were non-funerary, presumably cultic, seems plausible, especially considering the differences of vase shapes between those adorned with snakes and those without. Arguably, the gesture of holding hands represents dancing; moreover its appearance on a vessel did not denote whether the vase was funerary or religious, but instead this was left to both the shape of the vessel and the ancillary decorations present, such as serpents. It suggests that dancing was an important aspect of both funerary and religious events and that similar rituals occurred in both spheres.

If the gesture can be considered dancing, what is the meaning of it on a funerary vessel? As stated previously, none of the vessels depicting this dance are illustrated with representations of the deceased, thus suggesting this dance did not occur during the *prothesis* or *ekphora*. Instead, the dance could have occurred at one of various post burial rituals commonly performed by the deceased's family. It is generally believed these rituals concluded after 30 days with the *tria kostia*, "the ritual which concluded mourning held approximately one month after decease" (Garland 2001, 39). However, three further rituals are also known to have taken place within these 30 days, although the precise timing of these rituals remains unclear.

It is likely that the first ritual performed after burial was the *ta trita*, the third day ritual. The practice is known to have occurred on the third day, but it is unclear if the 'third' day represents three days after death, and therefore takes place during or immediately after the burial, or if the ritual occurred three days after interment (Kurtz and Boardman 1971, 144-146; Garland 2001, 40). It appears more likely that this event took place on the third day after death, marking the beginning of the mourning period. Alexiou argues that the thirtieth day rite is known to have occurred 30 days after death, not the burial, and therefore it seems unlikely that this would be different for the *ta trita* (Alexiou 1974, 7, 207-8, n.37 & 38). Similarly, Boardman and Kurtz also believe the event took place on the day of burial, suggesting the ritual was performed "to sow the earth[of the grave] with the fruits of its bounty," in order to bid farewell to the deceased (Kurtz and Boardman 1971:145). As to what the *ta trita* involved is also uncertain. Cicero, in his work *On the Laws,* may give some indication as to what occurred. He states, "when they had cast the earth over the dead, scattered the seeds of vegetables over the spot" (Cicero, *On the Laws* 2.63). Unfortunately, this remains the only possible description of the event.

The second ritual to have occurred appears to have also taken place on the day of the burial. The *perideipnon* consisted of a feast, where the "bereaved wore garlands and delivered eulogies on behalf of the dead" (Garland 2001, 39). The *perideipnon* appears to have taken the form of a modern day wake, with food a central part of the ritual. In Hegesippos' *Adelphoi*, a cook states, "Whenever I turn my talents to the *perideipnon*, as soon as they come back from the *ekphora*, all in black, I take the lid off the pot and make the mourners smile; such a tickle runs through their tums – it's just like being at a wedding"(Hegesippos, *Adelphoi* 11-16). This passage suggests that the practice not only took place on the day of the burial but that it also occurred in the house of the deceased. If Lucian's account in his work *Of Mourning* is correct, the feast was also a symbol of the end of three days of fasting (Lucian, *On Mourning* 24). The *perideipnon* does not appear to have been an iconographic motif drawn upon by the Athenians, with no extant representations of the practice known to the author.

The final ritual is the *ta enata* or the 9th day ritual. It appears to have taken place by the tomb where it has been suggested that "food, libations, and other offerings were placed on the new tomb" (Stears 2008, 142). Little is known about the rituals of the *ta enata* and the proposition that it involved libations and dedications remains speculative. However, it is perhaps possible that some of the representations found on later White Ground pottery depict this ritual.[3]

It is clear from the, albeit few, descriptions we have of post burial rituals that it is most likely the dancing women are

---

[3] For example, a White Ground lekythos in the Berlin Antikensammlung (inv.3372), represents women with short hair, indicating they could still be in the process of mourning.

depicted performing the *ta trita*. Cicero's description of the event clearly discusses the role of vegetal material, and with the branches, wreaths and 'upturned chevrons' held by the dancing women, it seems plausible that this is the event being depicted. Some scant archaeological evidence also suggests that burnt plant remains were present at the burial, as several Geometric inhumation graves in the Agora were discovered with "thick layers of ash with bits of animal bones, also burnt pots, also burnt pots which sometimes joined fragments of pots found in similar burnt deposits outside the grave"(Brann 1962, 112).[4] Whether the plant material was burnt during the *ta trita* is never mentioned, but the excavation of floral remains shows its importance within the burial process.

The *ta trita* was considered a crucial ritual in the burial process, as documented in Isaeus', *Menacles* 2.37: "In proof that I buried Menecles and performed the ceremonies on the third and ninth days and all the other rites connected with the burial". Although women appear to have been primary performers of the *ta trita*, men may have played a small part in the rituals as they appear on several vessels decorated with the dancing women. It is also highly likely that this ritual would have involved the singing of funeral dirges, which are believed to have been performed at a funeral (Vermeule 1979, 15). Ultimately, although Cicero's account provides some evidence that the dancing women may be part of the *ta trita*, it is impossible to be certain. Funerary rituals were rarely documented in detail by ancient writers and it is equally possible that the ritual depicted on the Hydriai represents an, as of yet, unknown practice.

As to why the depiction of funerary dancing disappeared on Attic funerary art, this may be a result of two independent factors. Firstly, following the popularity of funerary imagery during the second half of the 8th century, its depiction on Attic pottery significantly decreases after 700 BC. This period marks a rapid artistic revolution following the expansion of the Greek world into the wider Mediterranean. From approximately the last quarter of the 8th century BC, Athens enters the 'Orientalising Period', an age in which, as Hurwitt describes, "Near Eastern styles, techniques, decorative motifs, and even subject matter so pervade Greek bronzework, vase painting and sculpture" (Hurwitt 1985, 125). The dancing motif continues into the early Protoattic period, but disappears around the middle of the 7th century. It appears that with the rapid incorporation of Near Eastern motifs and painting styles, older iconography declined in popularity.

Secondly, following the scarcity of funerary imagery in the 7th century BC, the creation of the Black Figure style brought with it a return to mass produced funerary iconography. A result of this development was the production of Black Figure funerary plaques, or *pinakes*, and Black Figure funerary pottery, both of which were created between c.600 BC and approximately 500-475 BC. However, at the beginning of the 6th century BC, the Athenian lawmaker Solon instituted a series of austerity measures relating to funerary rituals. The exact dating of these laws is questionable, although there is some suggestion this may have occurred during his reign as archon in 594-593 BC (Garland 1989, 4). His exact laws were never recorded at the time and as such most of his edicts are known only from later sources. These laws limited the public display of emotion, for example by forcing the *ekphora* to take place before sunset and to travel through back streets whilst also limiting flagellation, the singing of prepared dirges and 'bewailing' at any of the funerary rituals (Plutarch, *Solon* 21).

As such, it is feasible that funerary dancing did occur at burials during the 8th and early 7th centuries BC. The disappearance of the motif from Protoattic pottery may be a result of the changing iconographic tastes of the Athenians. After 600 BC, when funerary imagery became popular again, the impact of Solonian funerary laws may have resulted in the abandonment of dancing at funerals along with other depictions of funerary rituals such as the *ekphora*. Alternatively, the event may have slowly been reintroduced into the funerary process, although iconographically it remained an unpopular subject.

Dancing on Black Figure Pottery

As mentioned previously, depictions of the *ekphora* and related rituals in Athenian art are relatively rare despite the important nature of the event. The *ekphora* is depicted several times during the Late Geometric period, although its use was short lived, appearing infrequently on Archaic and Classical funerary art.[5] However, the motif does appear on two Athenian Black Figure Kantharoi from the second half of the 6th century BC, painted by the so-called Perizoma Group.[6] On the first vessel, the deceased is being transported on a cart, whilst on the second, a group of men are illustrated carrying the body towards a grave. However, the most interesting aspect of the vases is that both are decorated with a group of dancing hoplites clearly following an *aulus* player (**fig. 5**). These Kantharoi provide the clearest representation of dancing within a funerary setting on all Athenian funerary art.

The dance being performed by the men is commonly known as the *pyrrhic*, an armed dance detailed in Plato's *Laws*. Plato describes the dance as "[representing] modes of eluding all kinds of blows and shots by swervings and duckings and side-leaps upward or crouching; and also the opposite kinds of motion, which lead to active postures of offence, when it strives to represent the movements

---

[4] This burning could be direct evidence of the giving of the fruits to the body performed in the *ta trita*, although Brann believes these were used for ritual feasting around the grave.

[5] One of the most detailed representations of the ekphora come from a small terracotta model found at Vari which dates to the second half of the seventh century BC. See Garland 2001, 32-33, fig. 9.
[6] A third Kantharos is known from the Basel Market, yet I will not discuss it here as I have not seen the vessel. Shapiro 2000, 326, 333-337. For the Perizoma Group see Shapiro 2000; Beazley 1956, 343-6 & Beazley 1971, 156-8.

Figure 5. Athenian Black Figure Kantharos from the Perizoma Group. Bibliothèque nationale de France Vase 355. (Photo provided with permission from the Bibliothèque nationale de France).

involved in shooting with bows or darts, and blows of every description" (Plato, *Laws* 815a). However, there seems to have been a variety of armed dances that could be performed in antiquity. These involved single warriors, to men staging mock battles, with several of these forms detailed in Xenophon's *Anabasis*. One of the dances described by Xenophon is remarkably similar to that depicted on the two Black Figure Kantharoi. Xenophon mentions that a group of Mantineans and Arcadians performed a dance, "arrayed in the finest arms and accoutrements they could command, and marched in time to the accompaniment of a flute playing the martial rhythm and sang the paean and danced, just as the Arcadians do in their festal processions in honour of the gods" (Xenophon, *Anabasis* 6.1.11).

However, despite the fact that the Kantharoi are clearly produced in an Athenian workshop, they appear to have been made specifically for the export market, with both vases discovered in Vulci, Etruria (Shapiro 2000; Osbourne 2010). Furthermore, the shape of the Kantharoi are heavily influenced by the one handled Kantharos, which was a popular Etruscan shape (Shapiro 1991, 633).[7] On one of the Kantharoi, the dress of the pyrrhic hoplites is also unusual, with the figures wearing a 'Perizoma', a loin cloth that appears to have been an attempt by the artist to dress the soldiers in attire common in Etruria (Shapiro 2000).[8]

---

[7] For more on the vessels see Poursat 1968, 550-615; Brommer 1989, 489; For Pyhrric dancing in Etruscan art, Camporeale 1987, 1-42.

[8] Shapiro also lists a variety of other key differences between pyrhhic dancers produced for the Athenian market compared to those on the Etruscan Kantharoi. Shapiro 2000, 336.

Music was frequently associated with Etruscan funerary practices, as demonstrated by an Aulos player standing next to a bier on an early 5th century BC *cippus* (Jannot 1988, fig. 5).More importantly, pyrrhic dancers regularly decorate both Etruscan cippi and funerary urns (Shapiro 2000, 335-336; Camporeale 1987; Jannot 1984, 331-338).

For the painter of the Kantharoi, decorating such a vessel with a pyrrhic dance would not be difficult. Armed dancing was a common enough iconographic motif on non-funerary pottery in Athens and its addition to a funerary vessel, although unusual, would not have been problematic (Poursat 1968).As such, the vessels, although fascinating, do not provide any evidence for the use of dancing in a funerary setting in Greece. As discussed previously, Solonian laws dictated that the *ekphora* should remain a relatively sombre affair and the addition of armed dancers to this ritual may have been inappropriate for a time of mourning.

Conclusion

Although the Black Figure Kantharoi and Geometric/Protoattic Hydriai are illustrated with dancing figures, it is obvious that the illustration of these motifs were not popular in Athens. Athenian artistic tastes favoured the more traditional *prothesis* scenes, common on pottery from mid-8th century BC to the end of the 4th century BC. Why this was the case is still uncertain, but it may have to do with importance of burial and remembrance. To not have the proper funerary rites was considered a great and grave dishonour to the deceased. Polynieces in *Oedipus at Colonus* begs not to be forgotten, "I beg you by the gods, leave me dishonoured, but give me burial and due funeral rites," (Sophocles, *Oedipus at Colonus* 1409-1415) whilst Electra laments the state of Agamemnon's burial stating "Agamemnon's grave, dishonoured, has not yet ever received any libations, or branch of myrtle, but his altar is barren of ornament" (Euripides, *Electra* 320-330). Sourvinou-Inwood states it best claiming, "the opposite of glory after death is to be forgotten" (Sourvinou-Inwood 1995, 376).[9] This urge to be remembered may have been facilitated iconographically in the representation of the *prothesis* scenes, the first event of the funeral and arguably the most important.

The relative lack of enthusiasm of portraying funerary rituals other than the *prothesis* is also observable within the written record. The overwhelming lack of descriptions of events such as the *perideipnon, ta trita, ta enata*, and *tria kostia* highlights the Athenians mindset that such rituals were not worth recording or portraying in any form of art. Instead, ancient writers group together these individual events as 'funeral rites'. This did not reduce their importance; it simply demonstrated that the rites were so well known by the Athenians it was not necessary to clarify their individual stages or specify their structure. As such, events that occurred at these stages of the funeral, such

---

[9] Her basis for this quote is a small fragment of poetry from Sappho (Fr. 55), where a wealthy woman is threatened with not being remembered after death.

as dancing, although important to the process, could have simply never been documented by Athenian artists.

The Geometric/Protoattic Hydria of Athens presents tantalising evidence that some form of dancing did occur during the funerary rituals of the Athenians. That they were part of the *ta trita*, or perhaps another event, is uncertain. Yet the lack of dancing in later Athenian funerary art is perhaps more apparent than real. With the singing of dirges being a major aspect of funerals, it seems highly unlikely that a form of ritualised dance did not occur at some stage of the funerary process. The line of the chorus of Herakles "what funeral dirge, or dance of death am I to raise?" even places dirges and dancing together, suggesting they are of equal importance. Ultimately, for the audience of the play, the 'dance of death' would have been a ritual that they would have known well, not from funerary iconography or literary sources, but from experience.

References

Ahlberg, G. 1971. *Prothesis and Ekphora in Greek Geometric Art*. Göteborg, P. Åströms.

Alexiou, M. 1974. *The Ritual Lament in Greek Tradition*. London, Cambridge University Press.

Beazley, J. D. 1956. *Attic black-figure vase-painters*. Oxford, Clarendon Press.

Beazley, J. D., 1971. *Paralipomena: Additions to Attic black-figure vase-painters and to Attic red-figure vase-painters (second edition)*. Oxford, Clarendon Press.

Brann, E. 1962. *The Athenian Agora Volume 8- Late Geometric and Protoattic Pottery*. Princeton, American School of Classical Studies.

Brommer, F. 1989. Antike Tanze. *Archäologischer Anzeiger*, 492-94.

Camporeale, G. 1987. La danza armata in Etruria. *Mélanges de l'Ecole française de Rome. Antiquité* 99(1), 11-42

Coldstream, J. N. 2003. *Geometric Greece: 900-700 BC*. New York, Routledge.

Garland, R. 2001. *The Greek Way of Death*. Ithaca, Cornell University Press.

Hurwit, J. M. 1985. *The Art and Culture of Early Greece, 1100-480 B.C*. Ithaca, Cornell University Press.

Jannot, J.-H. 1988. Musique et musiciens etrusques. *Comptes rendus des séances de l'Académie des inscriptions et belles-lettres* 132, 311-334.

Kurtz, D. C. and Boardman, J. 1971. *Greek Burial Customs*. London, Thames and Hudson.

Langdon, S. H. 2008. *Art and identity in Dark Age Greece, 1100--700 B.C.E*. Cambridge, Cambridge University Press.

Maidment, K. J. 1968. *Minor attic orators: I*. Cambridge, Harvard University Press.

Morris, I. 1987. *Burial and ancient society: The rise of the Greek city-state*. Cambridge, Cambridge University Press.

Osbourne, R. 2010. Why did Athenian pots appeal to the Etruscans? *World Archaeology* 33(2), 277-295.

Papadopoulos, J. K. 2007. *The art of antiquity: Piet de Jong and the Athenian Agora*. Princeton, American School of Classical Studies at Athens.

Parker, R. 1983. *Miasma: Pollution and purification in early Greek religion*. Oxford, Clarendon Press.

Poursat, J.-C. 1968. Les représentations de danse armée dans la céramique attique. *Bulletin De Correspondance Hellénique,* 92(2) 550-615.

Shapiro, H. A. 1991. The Iconography of Mourning in Athenian Art, *American Journal of Archaeology,* 95(4), 629-656.

Shapiro, H.A. 2000. Modest Athletes and Liberated Women: Etruscans on Attic Black figure vases. In B. Cohen (ed.), *Not the classical ideal: Athens and the construction of the other in Greek art,* 315-337 Leiden, Brill.

Sourvinou-Inwood, C. 1995. *"Reading" Greek Death: To the End of the Classical Period*. New York, Oxford University Press.

Stears, K. 2008. Death Becomes Her: Gender and Athenian Death Ritual. In A. Suter (ed.), *Lament: Studies in the ancient Mediterranean and beyond,* 139-155 Oxford, Oxford University Press.

Weiner, A. 1980. The Function of the Tragic Greek Chorus. *Theatre Journal* 32(2) 205-212.

Vermeule, E. 1979. *Aspects of Death in Early Greek Art and Poetry*. Berkeley, University of California.

Dancers' Representations and the Function of Dance in Han Dynasty (202 BC – 220 AD) Chinese Society

Marta Zuchowska

Institute of Archaeology
University of Warsaw

Abstract*: Dancers and dancing performances were among the most popular motifs in Han dynasty period funerary reliefs. Clay figurines and jade plaques representing dancers were also often placed in the graves of members of Chinese aristocracy. As we know from abundant literary sources, in earlier times dance in China was connected with ritual and ceremonial activities, but analysis of Han dynasty representations suggest that, in that time, it became merely entertainment. In this paper I would like to present some Han dynasty reliefs representing different categories of dance and their possible interpretations.*

Figure 1. Relief from Beizhai, Yinan County, Shandong province (after Zhang 2009, fig. 116).

Dancers and dance representations are amongst the most popular themes of Chinese art. They appear on prehistoric rock carvings, famous Zhou dynasty bronze vessels and tomb relief decorations, as well as providing a graceful subject for craftsmen producing terracotta figurines and jade pendants. Dance in ancient China was part of important state rituals as well as an element of popular culture and entertainment. Dance activities are known from the oldest literary sources (*Jiaguwen*, or oracle bone inscriptions). They were described by ancient historians and admired by Han and Tang dynasty poets.

In this paper, I would like to present some Han dynasty funerary reliefs and figurines and their possible interpretations.

Aristocratic tombs in ancient China were usually equipped with all the items believed necessary for the afterlife, in the form of real artifacts or models. In some graves, we find not only personal adornments and clothes, but also full sets of bronze and pottery vessels, weapons, musical instruments, medical equipment, etc. Models usually represented buildings and living creatures: granaries, parts of palaces (such as receptions or dancehalls), livestock and servants. Many graves are indeed eternal palaces full of masterpieces. Some of them were also decorated with reliefs, representing scenes of everyday life such as harvesting and wine production, or luxurious banquets with music, acrobats and dancers (Rawson 1999, 5-58).

These reliefs as well as all the objects usually placed in graves reflect the luxurious way of life of the noble class. Banquets with dancers represented on stones, along with figurines of dancer–servants, were intended to fulfill the necessity of entertainment in the afterlife, but at the same time show the social importance of this kind of activities to the upper class of society.

One of the most elaborate reliefs representing banqueting scenes was found in the village of Beizhai, Yinan county, in Shandong province (Zhang 2009, 168-169, fig. 116). Here we can observe the diverse dancing activities which were popular in Han dynasty China (**fig. 1**).

Starting from the left, in the lower part of the relief we can see a dancer and on his right side are seven plates and a small drum (**fig. 2**). These are items used for performing the so-called 'Seven Plates Dance', or *qipanwu*. According to ancient sources, this dance was usually performed by gracefully stepping between plates or quickly and lightly stepping on them (**fig. 3**). At the same time, the dancer was singing and beating the drum with their leg, creating the musical accompaniment to his performance. It is also possible that the sound of the plates touched by his legs

Figure 2. Qipanwu - *Seven Plates Dance. Relief from Zhengzhou, Henan province (after Zhang 2009, fig. 148)*

Figure 3. Qipanwu - *Seven Plates Dance. Relief from Peng county, Sichuan province (after Zhang & Liu 1996, fig. 6)*

Figure 4. Qipanwu - *Seven Plates Dance. Relief on the entrance door to the tomb M2, Zhejiang city (after Henan 1989: 4, fig 4).*

was also used as a part of the music. This dance is often represented in funerary art (**fig. 4**) and seems to have been very popular at court. It could be performed by male or female dancers and, according to iconographical sources, the number of drums and plates could change (Liu *et al.* 1983, 112 – 113).

The dancers were usually dressed in long sleeved apparel for *qipanwu* (**fig. 5**). This dress was also used for other kinds of dance. Sleeves were more than twice the length of the dancer's arms, so all movements created twirling movements of colourful cloth, which was one of the most characteristic features of Chinese dances. The so-called 'Long Sleeves Dance', or *changxiuwu*, was usually performed by one dancer with musical accompaniment from one or several instruments, and it was also a very popular motif in funerary art (**fig. 6**). It was represented not only in reliefs, but also in small plastics. Long sleeves dancer figurines were often present in the set of servant terracotta figurines, and in many graves small jade plaques (**fig. 7**) representing long sleeves female dancers were also found (Erickson 1994; Weifang 1993, 6, fig. 4).

Both *qipanwu* and *changxiuwu* are supposed to have originated and developed in the southern regions of China. They are often connected with the culture of the Chu state

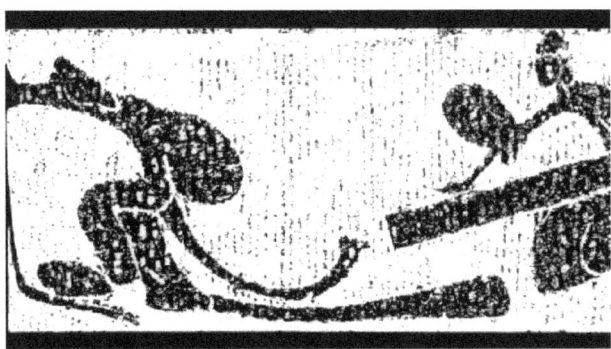

Figure 5. Changxiuwu – *Long sleeve dance. Relief from Anyang, Henan province (after Zhang 2009, fig. 149).*

Figure 6. Changxiuwu – *Long sleeve dance. Relief from Guanzhuang tomb, Mizhi county, Shanxi province (after Wu and Xue 1987, fig. 4)*

Figure 7. Jade plaque representing long-sleeve dancer, Changle county, Shandong province (after Weifang 1993, 6, fig. 4.4).

Figure 8. Jianguwu – *Dance with a standing drum. Relief from Beizhai, Yinan county, Shandong province (after Zhang 2009, fig. 116 detail).*

(mid XI c.- 223 BC), which at its height occupied territory across most of the south-eastern provinces of present-day China. The culture of this state was very rich and deeply influenced the subsequent Han dynasty musical and performing arts. According to the Chinese tradition, Han army was singing the Chu songs during the battle with the Chu soldiers, and even Gaozu, the first emperor of the Han dynasty, was fond of Chu dances and able to sing Chu songs (*Shiji* 8, Gaozubenji [Annals of the reign of Gaozu]).

On the left part of the Beizhai relief we can also see a man beating a large drum which stands on a pole, decorated by two umbrellas, long bands with plumes and a bird on the top **(fig. 8)**. This kind of drum was called a *jiangu* ('pole drum') and was usually used in a dance in which two dancers with drumsticks danced on both sides of the drum and beat it to provide the rhythm for their movements **(fig. 9)**. This dance was popular as a part of court festivities, described by Han dynasty poets. For example, Sima Xiangru in his *Zixufu* (*Master Nil*) mentions several court performances with drums. It is also often represented in funeral art **(fig. 10)**.

The *jiangu* dance (*jianguwu*) was not the only kind of dance performed with a drum. On a relief excavated in the tomb of Shanzi in Chulan, Su county, Anhui province, we can see 20 men in a line, all wearing long sleeved apparel **(fig. 11)**. Some of them hold a small round object with a handle in one hand. These objects are *pigu,* a kind of drum made from a piece of leather stretched on a round or oval frame and equipped with a handle. This drum and the dance associated with it were of non-Han provenience. It was popular among the Ba population, who lived on the southern border of the Chinese empire, and was incorporated, together with several other kinds of dance, into the Han dynasty's entertainment repertoire after the Ba population was defeated by Gaozu

Figure 9. Jianguwu – *Dance with a standing drum. Relief from Sancun, Tongshan county, Xuzhou city district, Jiangsu province (after Zhang 2009, fig. 112).*

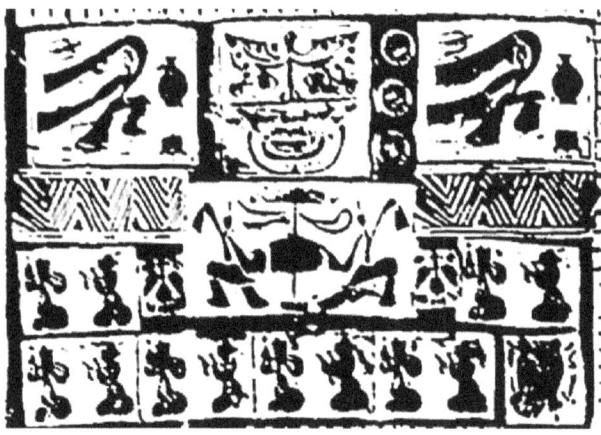

Figure 10. Jianguwu – *Dance with a standing drum. Relief on the entrance door to the tomb M2, Zhejiang city (after Henan 1989, 4, fig. 4 - detail).*

Figure 11. Piwu – *Dance with a Pi drum. Relief from Shanzi tomb, Chulan, Su county, Anhui province (fragment) (after Zhang 2009, fig. 146).*

(Liu *et al.* 1983, 111-112; Zhang 2009, 194). This dance lasted for a long time and is still a popular ethnic dance in the southern part of China; even today it can be seen during the famous New Year Parade in Hong Kong.

The other two dances often connected to the Ba population – a dance with a lance and a dance with a bow – began to become popular during the reign of Emperor Gaozu as part of *Juedixi*, martial art performances. *Juedixi* were organized at the court and had a martial as well as religious and ceremonial character. These performances gradually lost their religious and ceremonial connotations and started to become part of political strategy, especially in relation to barbarian envoys.

By the time of the Han emperor Wu (140-87 BCE), *juedixi* were not small meetings composed only of martial arts contests and performances, but large-scale festivities involving martial arts, music, dance, acrobatics and theatre-like performances showing a battle between a magician and a white tiger (McCurley 2005, 88-92).

In the middle of the *Beizhai* relief (**fig. 1**) we can see a procession of disguised dancers, wearing costumes of dragons, fish, leopards and phoenixes. The meaning of this dance is not clear; it seems it was also a part of the large-scale Han dynasty performances. Being a kind of theatre, these performances told stories connected with mythology or legends (Zhang 2009, 173).

The Jin dynasty chronicle *Jinshu* describes the rope-walking performance of the Han dynasty in these words: "During the reign of the Later Han dynasty, in the first day of the new year the emperor was facing the pavilion *Deyang* and was receiving congratulations from foreigners… Between two poles many *zhang* [measure of length: 3.3 m] long silk rope was fixed, there were two girls dancing on it, like on a cliff and they didn't fall down" (*Jinshu* 23: Yuezhi [Book of Music]).

In parts of the relief from Beizhai there are several acrobatic scenes. On the far left we can partially see a man playing with a sword over a few small balls, performing the so-called *Feijian Tiaowan* ('flying swords, jumping pellets'), then, slightly to the right, a larger man balances a cross-like construction on his head, on which three children perform acrobatics. In the middle of the scene, three people walk on a rope, while on the right side of the first plan we can see a cart with three horses in full gallop. On the cart is a high construction equipped with a large drum, a *jiangu*, on the top of which another man stands on his hands. Above this, two men display their horse-riding skills. Further examples of this kind of acrobatic performances can be seen on the reliefs from Anren, Daba county (**fig. 12**) and from Hanjiagu, Yishui county (**fig. 13**). These were also considered to be dancing activities and often accompanied other kinds of dancing and musical performances during large and important festivities.

Figure 12. Zaji – Acrobats performance. Relief from Anren, Daba county, Sichuan Province (after Zhang 2009, fig. 127).

Figure 13. Zaji – acrobats performance. Relief from Hanjiagu, Yishui county, Shandong province (after Zhang 2009, fig. 141).

Generally, the line between dance, acrobatics and martial arts during the Han dynasty was blurred. For example, the *qipan* dance is usually represented as part of a group, along with performing flying swords artists and jugglers. All activities represented on the Beizhai relief are included, in Chinese nomenclature, under the common term of *baixi,* which is usually translated as 'acrobatics'. Even if represented in a funerary context, they seem to be entertainment rather than ritual or ceremony. In fact it looks like most of the performing arts developed from old forms of ritual dances, but we rarely find dancers in ceremonial contexts represented in Han dynasty iconography. We can, however, find the ritual origins of the entertainments described above.

The long sleeves dance was represented in a ritual context before the Han dynasty reign. For example, on the Spring and Autumn period (770 – 476 B.C.) bronze bowl from Shanghai Museum, two long sleeves dancers take part in a large-scale ceremonial scene of ritual sacrifice (**fig. 14**). We also see them in the scene of offering on the *yi* vessel from the Eugene Fuller Memorial Collection, Seattle Museum (Weber 1966, fig.21c). It has been observed that these dancers could be priestesses or shamanesses (Erickson 1994) who, according to the *Chuci Zhaohun* and *Jiuge* ('Songs of Chu', 'Summons of the spirit', 'Nine songs'), had the power to summon spirits (*Chuciquanyi,* 32-54; 155-172).

Masked and costumed performances probably also originated from shamanic dances. It seems that during the Chu state, dancers wearing zoomorphic costumes performed dances which symbolized the expelling of demons (**fig. 15**). They were popular, especially in the area of the present-day Sichuan region; the relief scenes found here represent dancers wearing animal-like dresses as part of story-telling performances based on mythological themes (Zhang and Liu 1996, 95-101).

Finally, the use of the drum during the dance was connected with ceremonial and ritual dances from the beginning of Chinese culture. We find information about dances accompanied by drums on the oracle bone inscriptions from Shang dynasty times (Han 2008, 34); beating a drum was also one of the activities involved in the process of

Figure 14. Changxiuwu – Long sleeves dance in ritual context. Decoration of bronze bowl, Spring and Autumn period, Shanghai Museum (after Ma 1961, fig. 1).

Figure 15. Mask and costume performance (Expelling demon?) Relief on a stone box found in the rock-cliff grave in Xinjin, Sichuan province (after Zhang 2009, fig. 143).

summoning spirits described in the *Chuci Jiuge* ('Songs of Chu', 'Nine Songs') (*Chuciquanyi*, 32-54).

It is interesting that dances with weapons were virtually nonexistent in funerary representations during the Han dynasty. Dances performed with weapons known from literary sources obviously have some connection with Shang and Zhou dynasty ritual martial dances (*wuwu*). Reading Shang dynasty oracle bone inscriptions, we find that dance was one of the most important media for contacting the spirits.

Dance was also often considered a remedy for weather anomalies and, after consultation with an oracle, was performed to avoid such meteorological disasters as heavy rain sat an inappropriate time. In a large group of oracle-bone inscriptions from pit H3 in the Eastern Locus of Huayuanzhang, in the Yin ruins at Anyang, three types of dance are mentioned: Shang dances (*shangwu*), regional dances (*diyuwu*), and martial dances (*wuwu*). The latter was a dance performed with some kind of weapon (Han 2008, 34).

During the Zhou dynasty, the grand martial art performances, which developed into the later *juedixi*, were the autumn ceremonies organized with the aim of promoting *wu* (martiality). Autumn was associated with withering and death, and the ceremonies had a double purpose: as a religious ritual and as an entertainment for the watching crowds. Finally, they were also considered to be a good way of training martial skills during peacetime, under the control of the government.

During the Han dynasty, these ceremonies gradually lost their ceremonial function and seasonal character through the addition of regional dances with weapons and theatrical performances. They turned into large spectacles intended to delight spectators and impress foreigners. The later annalists ascribed this change to the Qin dynasty (McCurley 2005, 88-92).

Concluding this short sketch on Han dynasty dance performances, we can observe a deep secularization of dance during this epoch. At the beginning of Chinese civilization, dance was strictly connected with rituals and ceremonies and had its origin in the activities of the *wu*–shamans.

In Chinese mythology we find the story of Yu, the mythical founder of the Xia dynasty, who succeeded in subduing the waters which were threatening the country. As a result of this enormous effort, half his body was paralyzed, and Yu developed a strange way of walking called the 'step of Yu' (*yubu*). This became a sign that he could communicate with spirits, and we can conclude that Yu was a shaman. In later times, the 'step of Yu' was considered to be a way of healing. In the same way, shaman dancers, by the means of strange movements, were able to communicate with spirits and help in averting disasters (Mathieu 1989, 108-109; Erickson 1994, 52).

During the Shang dynasty, learning ritual dances was one of the duties of kings' sons and aristocratic children. During specified times of the year, these young boys had to learn and perform ritual dances under the supervision of the king. Skills such as knowledge about ceremonies, dance, singing, offerings, sacrifices and archery were believed to be the basis of education for young people, in order to prepare them for the role of high official or king (Han 2008, 33).

During the Zhou dynasty dance became more institutionalized, but was still an important part of ceremonies performed in the ancestral temple (Wu 1984, 80-81). During the Warring States period, rituals and cultic practices gradually lost their importance and dance activities became principally entertainment. Even if the knowledge about ritual sources of certain performances wasn't entirely forgotten, the state dance performances had symbolic and ceremonious rather than religious and ritual functions. The only image of the dancer that can be connected with ritual and shamanic activities is the long sleeves female dancer who, in a funerary context, could be interpreted as shamaness, summoning the spirits.

References

Chuci quanyi (Songs of Chu). 1991 Trans. Huang Shouqi and Mei Tongsheng. Guiyang, Guizhou renmin.

Jinshu, Fang Xuanling (History of the Jin Dynasty). 1999. Beijing, Zhonghua shuju.

Shiji, Sima Qian (Records of the Grand Historian). 1999. Beijing, Zhonghua shuju.

Erickson, S. N. 1994. "Twirling their long sleeves, they dance again and again…". Jade Plaque Sleeve Dancers of the Western Han Dynasty. *Ars Orientalis* 24, 39-63.

Han Jiangsu 2008. Cong Yinxuhua dong H3 buci pai pu kan shang daiwule (Studying Dance in Shang Dynasty Based on the Chronological Record of the H3 Oracle Bones of Eastern Locus at Huayuanzhuan in Yin Ruins). *Zhongguo Shiyanjiu* 2008:1, 21-40.

Henan (Henan Province, Institute of Cultural Relics) 1989. 'Zhejiangshi Nancang xijieliangzuo Hanmudefajue' (Excavations of Han Tombs on Nancang West Street in Zhengzhou city), *Huaxia Kaogu* 1989:4, 78-93.

Liu Zhiyuan, Yu Dezhang and Liu Wenjie 1983. Sichuan Handai huaxiang zhuan yu Handai shehui (Han dynasty pictorial bricks from Sichuan province and the Han society). Beijing, Wenwu Chubanshe.

Mathieu, R. 1989. *Anthologie des mythes et légendes de la Chine ancienne*. Paris, Gallimard.

Ma Chengyuan 1961. 'Mantan Zhangguo qingtongqi shangde huaxiang' (An informal discussion about the pictorial representations on the Warring States bronze vessels). *Wenwu* 10, 26-29.

McCurley, D. 2005. 'Juedixi'. An Entertainment of War in Early China'. *Asian Theatre Journal* 22.1, 87-106.

Rawson, J. 1999. The Eternal Palaces of the Western Han. The New View of the Universe. *Atribus Asiae* 59:1/2, 5-58.

Weber, C. D. 1966. Chinese Pictorial bronze vessels of the Late Chou Period. *Artibus Asiae* 28.4, 271-311.

Weifang (Weifang City Museum) 1993. 'Shandong Changle xian dongquan Han mu' (Excavation of Han dynasty tomb, found in Western Circle, Changle county, Shandong province). Kaogu 1993.6, 525-533.

Wu Zengde. 1984. *Handai huaxiangshi* (Han dynasty stone reliefs). Beijing, Wenwu Chubanshe.

Wu Lan and Xue Yong 1987. Shanxi Mizhixian Guanzhuang Dong Han huaxiangshimu (Excavations in the Eastern Han period tomb decorated with reliefs, found in Guanzhuang, Mizhi county, Shanxi province). *Kaogu* 1987:11, 997-1001.

Yao Xinzhong 2000. *An Introduction to Confucianism*. Cambridge, University Press.

Zhang Daoyi 2009, *Huaxiang shijianshang* (*Discussion on Stone Reliefs*). Chongqing Shi: Chongqing da xue chu ban she.

Zhang Xuezengand Liu Chuangang 1996. 'Qiantan Chu Wudaoyishu' (Brief talk on Chu state dancing art), *Huaxia Kaogu* 1996:2, 95-101.

Warrior Dance, Social Ordering and the Process of Polis Formation in Early Iron Age Crete

Anna Lucia D'Agata

Consiglio Nazionale delle Ricerche, Roma

Abstract: *Located at the northern end of the valley of Amari, within the territory of the Greek and Roman town of Sybrita, the settlement of Thronos Kephala is one of those new sites founded after the collapse of the territorial state system of the Late Bronze Age, at the beginning of the 12th century BC, which marked the dawn of a new era. This paper will focus on the scene of dance performed by three male warriors depicted on a PG clay krater found in the settlement, which provides a major contribution for the reconstruction of the social identity of the local early Greek community. The scene codifies the ways in which a ruling group in the community of Thronos Kephala assigned to itself the privilege of male initiation, and celebrated it, perhaps for the first time, with a ceremony which also included a banquet. This ceremony may be considered as the symbolic representation of the institution that enabled the society to ensure its continuation. Thus the vase from Thronos Kephala allow us to identify processes of social discrimination whose origins are closely bound up with the birth of communal institutions: processes which were set in motion to uphold the privileges of the ruling groups which emerged on Crete during the Early Iron Age and have to be considered as an important stage in the development of a secondary state formation.*

Male Dancing in Modern Crete

"Here in the church square, the constrast and tensions ... between official and Glendiot values come into sharp focus. The imposing church front ..., the community office, and the respect shown by men's removal of their headgear as they enter the church – these are testimonies to the official value. Pervading the same space however are the less obtrusive evidence of a contrary symbolic order: the surname clusters, the men's reluctance to enter the church at all, the agile male dancing with its implications of aggressive masculinity, and, by extension, of the skills of the good raider" (Herzfeld 1985, 67).

Thus Michael Herzfeld describes the conflicting social and political values in the pseudonymous village of Glendi, on the slopes of the Psiloritis massif in West-Central Crete, during the 1970s. Such a conflict appears to be tightly interwoven with the political vision shared by the villagers of the region: "They think of their community as a kind of small state, and thus assimilate the structures of statehood into an essentially segmentary view of the world" (Herzfeld 1985, xii). According to Herzfeld's ethnographic study, in traditional Cretan society male identity is constructed and maintained through specific performances or social actions, which take place in front of the public or are brought to life through story-telling. They consist of successive demonstrations of physical prowess, practical know-how, and deep insight. As well as animal rustling, drinking, song contests and card playing, dance is one of those social actions (Spencer 1985). Far from being simple symbolic performances, these actions are layered with multiple meanings and political and social conflicts: they ultimately

Figure 1. Warrior dance krater THK02/1 from the settlement of Thronos Kephala (photograph by M. Ierman).

resolve the network of binary oppositions of 'them' and 'us', which range from the local to the global level, and inform and pervade the traditional society which is still alive and well in this part of the island.

Herzfeld's approach to modern Cretan male dancing may represent an appropriate sociological background for the interpretation of the warrior dance scene depicted on a clay krater (**fig. 1**) found at Thronos Kephala (ancient Sybrita) (**fig. 2**), a settlement of the Early Iron Age on a

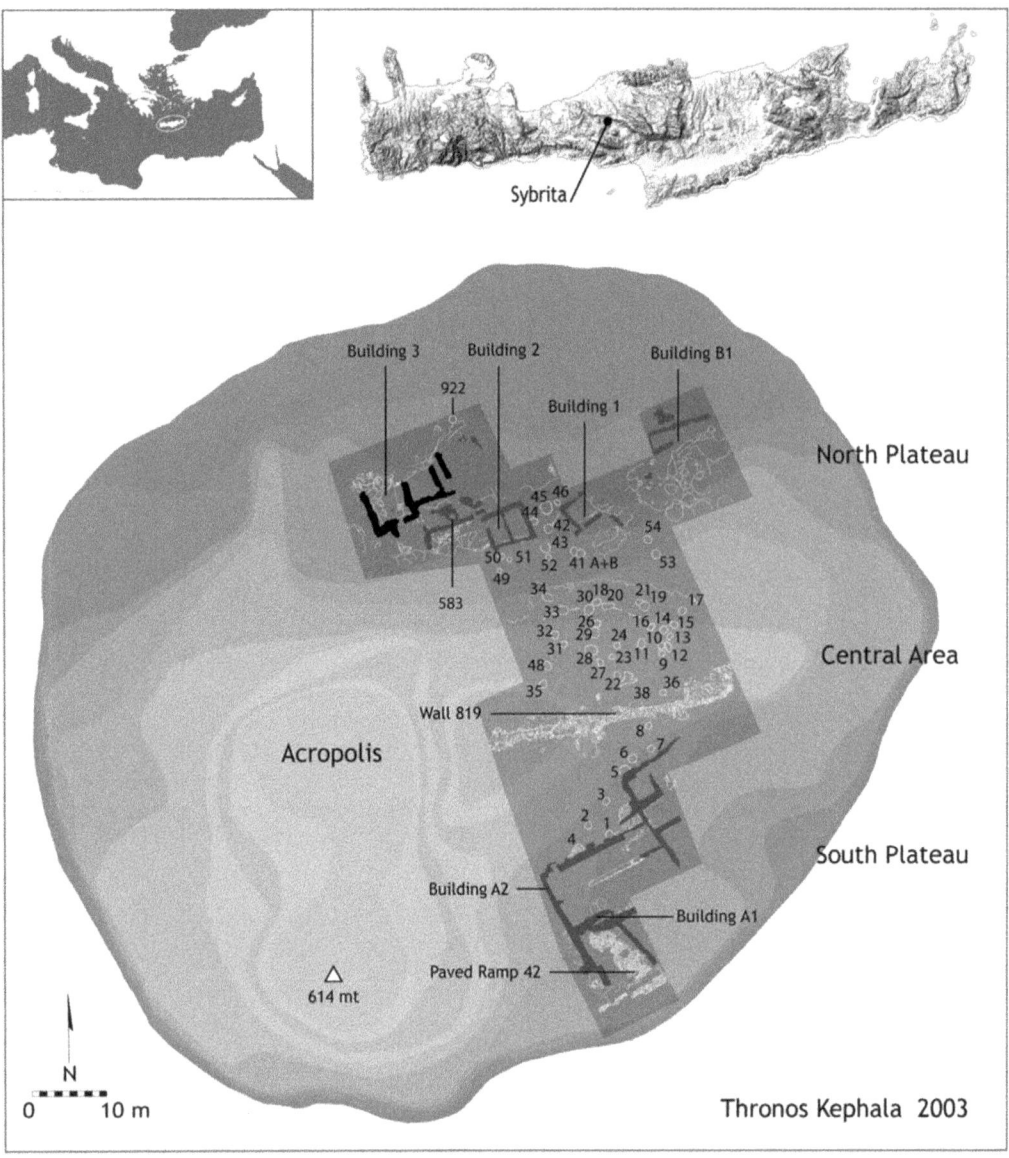

Figure 2. Map of Crete with the indication of Sybrita, and plan of the settlement of Thronos Kephala at the end of the 2003 excavation season (digital image processing by A. Di Renzoni).

hill at the western end of the Psiloritis massif (Rocchetti 1994; D'Agata 1999; D'Agata 1997-2000; D'Agata 2008; D'Agata forthcoming a). This is the steep region described by Herzfeld which includes the highest peak of Crete, Mount Ida. This is also the most conservative region of the island, where Greek mythology located the birth of Zeus in a cave where young Cretan warriors, the Kouretes, drowned out the cries of the infant god by performing a noisy war-dance to defend him from his father (Willets 1962, 98-100, 211-217). Here, modern archaeology has identified the Idean cave as the legendary birthplace of Zeus and the only pan-Cretan sanctuary in both Greek and Roman times (Sakellarakis 1988; Sporn 2002, 218-223). Even today, the Psiloritis region and the valley of Amari maintain a strong identity, founded on a pastoral economy and the existence of the feuding society described by Herzfeld, whose values are based on the absolute preeminence of masculinity.

Located at the northern end of the valley of Amari, within the territory of the Greek and Roman town of Sybrita, the settlement of Thronos Kephala (**fig. 2**) is one of those new sites founded at the beginning of the 12th century BC which, after the collapse of the territorial state system of the Late Bronze Age, marked the dawn of a new era. Like Sybrita, many of these sites were destined to become important Cretan towns of the Archaic and Classical age.

The research on the settlement of Thronos Kephala aims to reconstruct the local social and political processes which developed between the 12th and the 7th centuries BC. Among its goals is also an enquiry into how settlements like the one on Kephala, humble heirs of a state-level society, were able to transmit some of the achievements of the Bronze Age and contribute to the shaping of the social and political complexity of Archaic Greece. The warrior dance

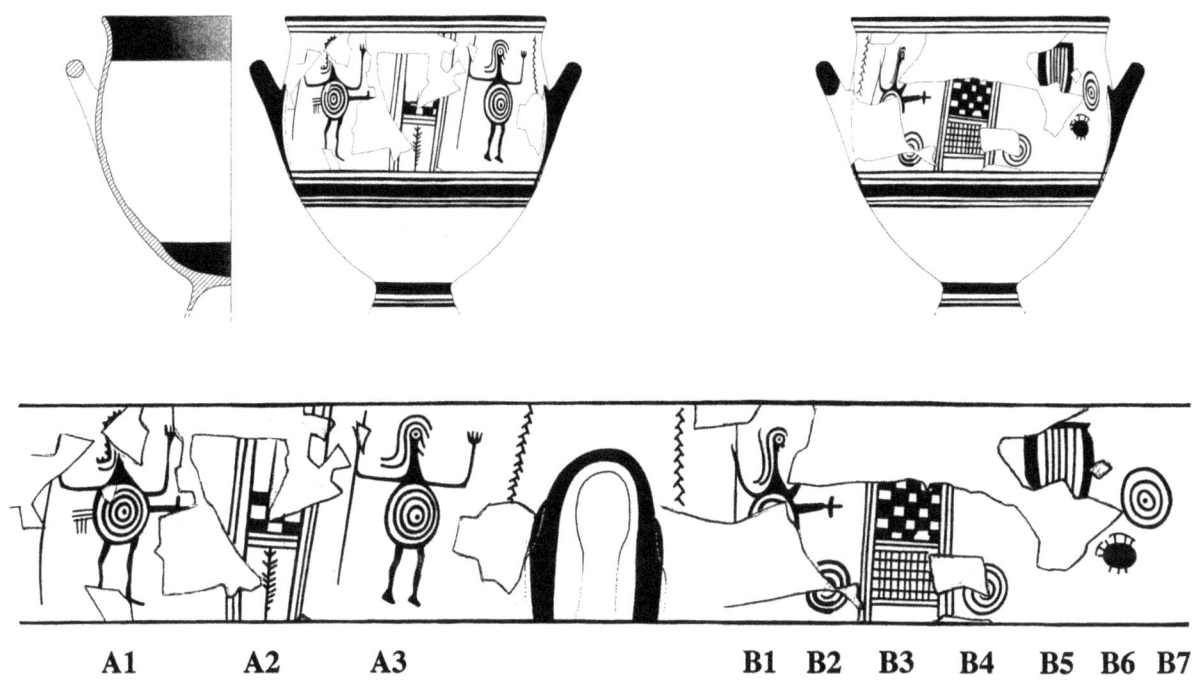

Figure 3. The warrior dance krater from Thronos Kephala, scale 1:3 (drawings by G. Merlatti).

krater found in the course of the 2002 excavation campaign provides a major contribution to the reconstruction of the social identity of the local early Greek community. The final publication of this vase may be found in D'Agata 2012. In this article I shall focus on the warrior dance which is depicted on it.

However rare, scenes of dance were represented on clay vases in the Aegean in the course of the Bronze age until the 12th century BC (German 2005, 53-71; Aamont in this volume) and, again, in the Geometric period (Tölle 1964, 77-79; Kaufman-Samaras 1972; Rombos 1988, 137-138; Langdon 2008, 144). The scene on the krater from Thronos Kephala dates from the Early Protogeometric, whose traditional chronology roughly covers the 10th century BC (Coldstream, Eiring and Forster 2001, 22; but see Weninger and Jung 2009). It constitutes a unique piece of evidence and requires a reconstruction of the significance of the performed action and of the relevant social context.

The Warrior dance Krater from Thronos Kephala

The figured scenes

The krater from Thronos Kephala displays on its sides a painted, figured scene. Side A depicts two armed warriors at the sides of a central panel, while side B shows a warrior and a group of objects, again at the sides of a central panel (**fig. 3**). Armed warriors are a common theme in the pictorial pottery of Mainland Greece in the 12th century BC, with the appearance of a quantity of vessels, especially kraters, decorated with warriors, fighting or in procession, celebrating belligerence and warfare (Vermeule and Karageorghis 1983, chapter XI; Deger-Jalkotzy 1999; Crouwel 2006, 17-18; Dakoronia 2006; Deger-Jalkotzy 2006, 174-75; Crouwel 2007, 74-75; Deger-Jalkotzy 2008). To find subsequent examples of figured scenes we have to take a leap forward to the 9th century BC and to Knossos to find a series of vessels – mainly kraters – decorated with figured scenes characterized by a mixing of themes of Near-Eastern origin with others of Minoan derivation (Coldstream 1980; Coldstream 1984; Coldstream 1991; Coldstream 2006; Stansbury-O'Donnell 2006).

In comparison with the examples cited above, the Thronos Kephala krater shows quite striking differences. First, it dates to the Early Protogeometric. Second, the subject depicted cannot be described as fantastic, nor can it be traced back to the Minoan tradition or Oriental influence. Rather, structural and contextual analysis of the scenes painted on the vessel show that they have to be seen as the manifestation of a specific social and cultural context.

The Thronos Kephala krater exhibits a decorative formula consisting of motives which also include concentric circles on both sides of a central panel, a formula already known on several Early Protogeometric kraters from the North Cemetery at Knossos (Coldstream, Eiring and Forster 2001, 47, fig. 1.13, b-c). Here, however, a major innovation appears: the decorative motives have been replaced by warrior figures (**fig. 3**: A1, A3, B1), while in one case the area has been filled with a group of musical instruments (**fig. 3**: B5, B6, B7). Among them are a lyre (B5) and its sound box (B6), which, as usual in the ancient world, is a tortoise shell. Close to them is what appears to be a tympanum (B7) – a circular drum of ox skin, created in imitation of the circular shields – represented elsewhere on the vase (**fig. 3**: B2, B4). Lyre and tympanum are indicators of the musical context in which we imagine the warrior dance was performed (D'Agata 2012).

The warrior figures are shown as full silhouettes, standing well above the base line. A1 wears a crested helmet, a circular shield covers his chest, and a sword, the end of which emerges on his left hip, hangs from his side. He also holds a spear in his right hand; his left arm is raised high in the air, with the palm of his hand open. The other two warrior figures closely resemble A1. A3 carries a shield and spear, but has neither crested helmet nor sword. B1 is the figure with most missing: he bears shield, spear and sword but no helmet. The three figures are not shown in the act of fighting, as implied by the vertically-held spear, nor advancing in procession, as inferred by the feet set obliquely well above the base line. The warriors are in fact shown in the act of leaping to the right. In the Late Geometric figurative repertoire, figures with at least one arm raised and the palm of the hand open, and in some cases the feet taking off from the base line, are interpreted as dancing (Tölle 1964, 77-79; Kaufman-Samaras 1972; Rombos 1988, 137-138; Ceccarelli 1998, 16). The armed warriors depicted on the Thronos Kephala krater may then be considered to show the act of dancing or, in other words, the performance of an armed dance. This interpretation also accounts for the presence of musical instruments on side B.

Judging from the rarity of metal panoplies in tombs, settlements or sanctuaries at the end of the second millennium BC, the scene on the vase from Thronos Kephla surely implies that the display of weapons, even in a figured scene, allude to the wealth and rank of their owners (Snodgrass 1964; van Wees 1998; Everson 2004). Thus the krater should be considered as an object of prestige that circulated among the members of the most prominent groups of the local community. In addition, this scene should also be considered as the earliest representation of armed dance, or pyrrhic dance, in early Greece. Its appearance implies a clear shift from the war symbolism of the scenes depicted on LH IIIC kraters: represented here are not warriors marching in procession, or fighting, but individuals who, like Homer's heroes, wear weapons as a mark of distinction.

Dance, and in particular armed dance, cannot be considered as an action of everyday life, but must have been performed on particular occasions, and is therefore to be attributed with ritual significance. Thus the next question we have to tackle is: what prompted the birth of this type of representation, and what value was ascribed to it by those who had the opportunity to view or hold the warrior dance krater in their hands.

The archaeological context

The krater was found in the settlement of Thronos Kephala, in Building 3 (**fig. 1**) which was constructed in Subminoan II (D'Agata 2007; D'Agata 2011) – during the 11th century BC– and suffered its first severe destruction in the Early Protogeometric period. The vase was found on the floor level of the west room, and the materials that came to light with it are typical of a Cretan banqueting set from the 10th century BC: kraters, skyphoi and cups together with vases which do not usually occur in a domestic context, such as an askos and an amphoriskos (**fig. 4**) used for storing/pouring liquids of some value, perhaps perfumed oils or condiments. This suggests that the warrior krater was used in the context of convivial gatherings of what we could call 'a local aristocracy', and, going further, given its complex figured representation, the warrior dance krater may have been produced for a specific event connected to a banquet, an event considered memorable at least at the local level.

The three warriors painted on the vase are shown moving towards the right, in one main direction, one behind the other, or in a circular motion. Incidentally, it should be noted that on Greek painted vases heroes or victors always move towards the right (Stansbury-O'Donnell 1999, 82). If the figures are moving in circular motion, we should assume that the ideal centre of the action was the central panel on the two sides of the vase (**fig. 3**: A2 and B3), implying that it represents two diverse faces of the same built structure, i.e. an altar or a building.

When the Thronos Kephala vase was painted the Greek alphabet had yet to make its appearance. Tales of heroes, in a narrative form, had certainly been circulating in Greece for presumably many centuries, but the representation on our vessel finds no comparison in the Late Bronze Age repertoire. For the time being we must consider it an original invention of a craftsman who must have found inspiration in the milieu in which he lived. If this scene represents a specific action performed in a ritual context, we must conclude that the images on the Thronos Kephala krater came about as a response to an impulse from the social world to which the craftsman belonged, and evidently connected with the function served by our vessel in its original context. It follows that we have to reconstruct the social context within which the action represented – the armed dance – takes its place. Given that we have no contemporary sources for the artefact in question – neither iconographic nor literary – this will be done in the light of the later literary and iconographic sources.

Warrior dance and the Cretan System of Education

In the ancient Greek world the armed dance, or pyrrhic, was a form of symbolic representation where the emphasis is placed on its war-waging potential. In none of the myths on the origin of the armed dance is there an opponent to face: rather, the dance is performed to demonstrate personal ability. However, the significance of the pyrrhic reaches well beyond the military sphere, as the myths connected with the origins of the dance itself indicate.

According to Ceccarelli (1998), it played a primary role above all in youth education. As indicated in a passage by Homer (*Il.* 7.237-241), the young appear to have been

Figure 4. Askos THK02/65 and amphoriskos THK04/43 from the settlement of Thronos Kephala (photographs by M. Ierman).

initiated in the art of war with paramilitary exercises accompanied by music. So it was that they learnt to handle shield and spear, and acquire the agility needed in combat to fight side by side with fellow warriors. In the Cretan system the young, up to 17 years of age, were gathered into groups to receive instruction for adult life under the general supervision of the adult males of the community. They were then promoted to full citizenship. To mark this transition an annual ceremony was performed during which the young were wedded and presented with their armour. Thus symbolically the death of the youth and the birth of the new – armed – citizen was sanctioned (Brelich 1969, 196-207; Lebessi 1991, 103-13; Chaniotis 1996, 21 and n. 83; Willets 1955, 15-17 and 120-123; Willets 1962, 46-53; Willets 1965, 112-18).

A precise description of the dance is given by Plato in the *Laws* (815a). Typical movements are to leap abruptly in the air and crouch – movements that did not befit a phalanx of hoplites so much as small bands of warriors engaged in an ambush, representing a type of guerrilla warfare generally associated with the city of Sparta. Thus the pyrrhic is a sequence of movements rehearsing the military manoeuvre typical of an ambush.

In the Classical age the pyrrhic dancers were nude and bore shield, helmet, and a weapon, or they would imitate with their hands the wielding of sword, spear or javelin. While the shield was a constant in the equipment of the pyrrhic dancers, the weapon could vary, just as it does on the warrior dance krater.

The two areas from where the dance was thought to have originated were Crete and Sparta, and local legends attribute the pyrrhic to geographically-localised heroes and divinities. On Crete, they were the Kouretes. In the *Bacchae* (120-34), Euripides describes how the Kouretes created a *byrsotonon kykloma*, or drum of tightened hide, in a sacred cave of Crete 'with the aim of drowning the cries of the infant Zeus, lest his father Chronos should find and devour him' (Dodds 1960, 83). Otherwise known as Corybants, the Kouretes were semi-divine figures attributed with the invention of both the armed dance and the tympanum. As originally suggested by Jane Harrison (1908-1909), the Kouretes may have represented the model for the initiation of adolescents in groups, the symbols of the initiated young Cretans.

When, one might ask at this point, were the young Cretans called upon to perform the armed dance? An extraordinary epigraphical Cretan document – the hymn to the Greatest Kouros of Palaikastro – offers a possible context. The inscription was found in the area of the Temple of Zeus at Palaikastro, within the territory of the polis of Itanos, on the far eastern coast of Crete. This is in fact a Roman copy of an original attributed to the 4th or 3rd century BC, and probably derived in turn from a more ancient version. The hymn (*IC* III, II.2) recounts the myth of the birth and childhood of a nature divinity identified with Zeus Dictaeus. In the 5th and 6th strophes the chorus call upon the god to endow the flocks, cattle, and crops with fertility, and the young citizens entrusted with the protection of the community. The verb used to convey the concept of fertilising is *trosko eis,* meaning 'to jump in', 'impregnate', and by extension 'fertilise', implying an action much like the pyrrhic leap or sudden emergence from a bush. In the refrain the god is invoked as Kouros, son of Kronos, guide to the daimones, and invoking him are the young initiates of Itanos. Accompanied by the sound of flute and lyre, they dance in a circle round the altar of the God, as did the Kouretes armed with shields about the baby Zeus.

The hymn was probably sung at the annual feast of Zeus, which was also the occasion when the new citizens of Itanos swore their oath of allegiance to the polis. Closely connected with the ephebi of Itanos, the hymn to the Greatest Kouros thus suggests that both the armed dance and the ephebi's oath took place during the same ceremony (Perlman 1995).

Created to imitate the rapid warrior's shift into and out of the battle-field, the armed dance was concluded, according to Ceccarelli (1998, 217), by the integration – or reintegration – of the group of dancers into the collective. Moreover, on a psychological level, it may also be interpreted as a dance of transition from one state of mind to another, and a discipline meant to educate, control and coordinate youthful aggression (Stehle 2000).

The hypothesis that initiation rites should be seen as the necessary starting point of many of the Greek myths (Jeanmaire 1939; Brelich 1969; Calame 1977) has recently been challenged, and it has been shown that mainland Greece actually had very few initiation rituals (Price 1999, 17; Stehle 2000; Graf 2003; Iles Johnston 2003, 155). Consequently, the armed dance cannot be considered to be exclusively linked to similar rites. However, the hymn of Palaikastro and the archaeological evidence from the sanctuary of Kato Symi (Lebessi 1985; 2002; Marinatos 2003) confirm the existence of relevant initiation rites on the island from an early date, and the important role that these rites played in the formation of some of the social and political institutions that were to be typical of the Archaic age. In other words, the evidence available from Crete legitimises the interpretation of the scene on the krater from Thronos Kephala as a reflection of an initiatory practice of young males to warrior status.

The Warrior Dance and the Process of Polis Formation in Central-Western Crete

It is common practice to trace the origins of many of the institutions typical of Archaic Crete back to Minoan Neopalatial society, even though the problem of how this influence was transmitted over the lengthy period stretching from the mid-Second millennium to the first attestations of the Archaic period has never been faced. Recently, however, a contrasting reconstruction has shown how the 12th century BC constituted a clear social turning point: it saw the rise of the Cretan system of educating youth, which was to be one of the most significant and characteristic institutions of the island's Archaic poleis (Lebessi 2002). In other words, a clear divide between Minoan society and Archaic social structure has been convincingly proposed, with this new reality being formed during the Dark Ages.

During the 12th century BC on Crete, as well as on mainland Greece, a new importance was assigned to the representation of the male figure as a warrior. We have already mentioned the Late Helladic IIIC kraters which depict warriors in arms, marching or fighting, and celebrating the art of war. Another recurrent subject on Cretan vessels dating from the 12th to the 9th century BC – i.e. the krater from Mouliana (D'Agata 2007, fig. 13.1) as well as several kraters from Knossos (Coldstream, Eiring, and Forster 2001, 49, fig. 1.14, b-c), all of them employed in funerary contexts – are hunting scenes. Being an activity proper of an aristocratic lifestyle since at least the Mycenaean age, hunting scenes displayed during banquets or funerals of prominent individuals helped reaffirm the ideology connected to the masculinity of Cretan ruling groups during the Early Iron Age.

The association between the lyre and warriors on the krater from Thronos Kephala is already attested on a Cypriot kalathos from Kouklia, of the 11th century BC, where a warrior armed with a sword is represented in the act of playing the lyre (Vermeule and Karageorghis 1982, 129 XI.39 and 136 XI.59; Iacovou 1988, 18 no. 29 and 49, figs. 68-71). The two vases, from Thronos and Kouklia, evoke figures (the warrior) and social customs (the use of music on convivial occasions) that have certain aspects in common. Nevertheless, the Cypriot citharist, depicted out of context, may refer to one of the many social occasions upon which the local ruling group would make use of music (Deger-Jalkotzy 1994, 22-23; Knapp 2008, 283): a banquet, a funeral, or perhaps the narration of traditional tales. This practice is coherent with a court system and therefore with one of the new political formations which emerged in the 11th century BC in Cyprus and would eventually result in the city kingdoms of the later Iron Age.

By contrast, the scene depicted on the krater from Thronos Kephala documents a paradigm shift in the social ideals connected with manhood. It shows a ritual action founded on an armed dance, which may have involved the acquisition of a full armour in occasion of a ceremony of male initiation. If the krater was the defining shape and symbol of the banquet, the scene depicted codifies the ways in which a group in the community of Thronos Kephala assigned to itself the privilege of male initiation, and celebrated it, perhaps for the first time, with a ceremony which also included a banquet. This may be considered as the iconographic and ritual expression of the institution that enabled the society to ensure its continuation. It was in fact the prototype of an institution typical of the aristocratic system which was to dominate the peculiar form of the Cretan polis: a system based on age-groups in which kinship played a major role (Willets 1955; 1965, 56-75; Whitley 2000, 252).

Thus, the warrior dance krater seems to express an ideal of manhood which goes beyond the celebration of physical abilities. It seems to imply the 'domesticated warrior' evoked by Langdon (2008, 250): a leadership which wants to assure the continuity of its household and calls for models of stability that could be connected with the formation stage of civic institutions within an on-going process of state-formation. In other words, we may conclude that at least since the 10th century BC in some Cretan communities the warrior dance was performed, and even represented on clay vases, to play symbolic and practical roles which would not have been dissimilar from those it played in the Greek Geometric age. The armed dance may imply the consolidation of the ruling groups which emerged after the collapse of the political state system of the Late Bronze Age and can be interpreted as one of the many indexes of a process of state formation which are evident at some Cretan sites of the 10th century BC. Here I refer to the nucleation of population and the formation of large sites, especially in Central Crete (Wallace 2006, 641-647); to an increase in the complexity of funerary rites and to the transformation of the religious system (D'Agata 2006); to the reuse of the monumental ruins of the Bronze Age in terms of territorial legitimization (D'Agata 2006); and to the development of stronger links with the Near East. These sites were designed as protopoleis (Wallace 2006, 648; Wallace 2010); however their political structure remains obscure, still pending clear archaeological evidence able to support a reconstruction of their political institutions. Thus, political communities could have preceded the formation of the polis as an urban phenomenon (Yoffee 1997, 261-62; Whitley 2000, 168; Hall 2007, 67-70).

In conclusion, the armed dance represented on the krater from Thronos Kephala shows the existence of an initiatory rite reserved to young males to be connected to ruling groups of aristocratic type. One of the most significant features of the Cretan poleis in the Archaic period is the creation of new political institutions based on the recognition of privilege for some social groupings, notably the *syssitia,* or common meals, and male initiation rites (Willets 1965, 86-87, 111-123; Guizzi 1997). Hence we can identify processes of social discrimination whose origins are closely bound up with the birth of these institutions, processes which were set in motion to uphold the privileges of the ruling groups which emerged during the Early Iron Age. An 'aristocratic' group became recognisable during the 10th century BC at Thronos Kephala, a period that can be seen as a turning point for many Cretan communities, paving the way for the sort of increasing complexity and integration of corporate groups (van der Vliet 2011) that marks the emergence of the Greek city-states.

References

Bosanquet, R. C. 1908-1909. The Palaikastro hymn of the Kouretes. *Annual of the British School at Athens* 15, 346-356.

Brelich, A. 1969. *Paides e parthenoi* (IG 36). Rome, Edizioni dell'Ateneo.

Calame, C. 1977. *Les Chœurs de jeunes filles en Grèce archaïque.* Rome, Edizioni dell'Ateneo and Bizzarri.

Ceccarelli, P. 1998. *La pirrica nell'antichità greco romana. Studi sulla danza armata.* Pisa and Rome, Istituti editoriali e poligrafici internazionali.

Chaniotis, A. 1996. *Die Verträge zwischen kretischen Städten in der hellenistischen Zeit.* Stuttgart, Franz Steiner Verlag.

Coldstream, J. N. 1980. Knossian Figured Scenes of the Ninth Century B.C. In Πεπραγμένα Δ' Διεθνούς Κρητολογικού Συνεδρίου, Ηράκλειο 29 Αυγούστου-3 Σεπτεμβρίου 1976, 67-73. Athens, University of Crete.

Coldstream, J. N. 1984. A Protogeometric Nature Goddess from Knossos. *Bulletin of the Institute of Classical Studies* 31, 93-104.

Coldstream, J. N. 1988. Some Minoan reflexions in Cretan Geometric art, in J. H. Betts, J.T. Hooker and J. R. Green (eds.), *Studies in honour of T. B. L. Webster,* 23-32. Bristol, Classical Press.

Coldstream, J. N. 2006. The long, pictureless hiatus. Some thoughts on Greek figured art between Mycenaean pictorial and Attic Geometric', in E. Rystedt and B. Wells (eds.), *Pictorial Pursuits: Figurative Painting on Mycenaean and Geometric Pottery* (Acta Instituti Atheniensis Regni Sueciae in 4°, 53), 159-163. Stockholm, Svenska Institutet i Athen.

Coldstream, J. N., Eiring L. J., and Forster, G. (eds.) 2001. *Knossos Pottery Handbook. Greek and Roman* (British School at Athens Studies 7). London, British School at Athens.

Crouwel, J. H. 2006. Late Mycenaean Pictorial Pottery: A brief review, in E. Rystedt and B. Wells (eds.), *Pictorial Pursuits: Figurative Painting on Mycenaean and Geometric Pottery* (Acta Instituti Atheniensis Regni Sueciae in 4°, 53), 15-22. Stockholm, Svenska Institutet i Athen.

Crouwel, J. H. 2007. Pictorial pottery of LH IIIC middle and its antecedents, in S. Deger-Jalkotzy and M. Zavadil (eds.), *Chronology and Synchronisms II. LH IIIC Middle. Proceedings of the International Workshop held at the Austrian Academy of Sciences at Vienna, October 29th and 30th, 2004,* 74-88. Vienna, Verlag der Österreichischen Akademie der Wissenschaften.

D'Agata, A. L. 1999. Defining a pattern of continuity during the Dark Age in central-western Crete: Ceramic evidence from the settlement of Thronos/Kephala (ancient Sybrita). *Studi Micenei ed Egeo-Anatolici* 41/2, 181-218.

D'Agata, A. L. 1997-2000. Ritual and rubbish in Dark Age Crete: The settlement of Thronos/Kephala (ancient Sybrita) and the pre-classical roots of a Greek city. *Aegean Archaeology* 4, 45-59.

D'Agata, A. L. 2006. The cult activity on Crete in the Early Dark Age. Changes, continuities, and the development of a 'Greek' cult system, in S. Deger-Jalkotzy and I. Lemos (eds.), *Ancient Greece. From the Mycenaean Palaces to the Age of Homer,* 397-414. Edinburgh, University Press.

D'Agata, A. L. 2007. Evolutionary Paradigms and LM III. On a Definition of LM IIIC, in S. Deger-Jalkotzy and M. Zavadil (eds.), *LH IIIC Chronology and Synchronism II. LH IIIC Middle. Proceedings of the International Workshop held at the Austrian Academy of Sciences at Vienna, October 29th and 30th, 2004,* 89-118. Vienna, Verlag der Österreichischen Akademie der Wissenschaften.

D'Agata, A. L. 2008. Rito, ambiente e paesaggio a Creta nel XII secolo a.C. La fondazione dell'insediamento di Sybrita, paper presented at the meeting *Il territorio e gli insediamenti in Europa e nel Mediterraneo: un Progetto CNR tra approcci metodologici e prospettive*, Roma, 29-30 aprile 2008.

D'Agata, A. L 2011. Subminoan: A neglected phase of the Cretan pottery sequence, in W. Gauss, M. Lindblom, R. A. K. Smith, J. C. Wright (eds.), *Our Cups are Full. Pottery and Society in the Aegean Bronze Age*, 51-64. Oxford, BAR Publishing.

D'Agata, A. L. 2012. The power of images. A figured krater from Thronos Kephala (ancient Sybrita) and the process of polis formation in west-central Crete. *Studi Micenei ed Egeo-Anatolici* 54, forthcoming.

D'Agata, A. L. forthcoming a. *Thronos Kephala (antica Sybrita): le fosse rituali in prossimità dell'insediamento nell'area centrale* (IG). Rome, Istituto di studi sulle civiltà dell'Egeo e del Vicino Oriente, CNR.

D'Agata, A. L., and Karamaliki, N. 2000. Campagna di scavo 2000 a Thronos/Kephala (Creta, Grecia). *Studi Micenei ed Egeo-Anatolici* 42/2, 337-340.

D'Agata, A. L., and Karamaliki, N. 2002. Campagna di scavo 2002 a Thronos/Kephala (Creta, Grecia). *Studi Micenei ed Egeo-Anatolici* 44/2, 347-355.

Dakoronia, F. 2006. Mycenaean pictorial style at Kynos, east Lokris, in E. Rystedt and B.Wells (eds.), *Pictorial Pursuits: Figurative Painting on Mycenaean and Geometric Pottery* (Acta Instituti Atheniensis Regni Sueciae in 4°, 53), 23-29. Stockholm, Svenska Institutet i Athen.

Deger-Jalkotzy, S. 1999. Military Prowess and Social Status. In R. Laffineur (ed.), *POLEMOS. Le contexte guerrier en Égée à l'Âge du Bronze* (Aegaeum 19), 121-131. Liège and Austin, Université de Liège, Historire de l'art et archéologie de la Grèce antique and University of Texas at Austin, Program in Aegean Scripts and Prehistory.

Deger-Jalkotzy, S. 2006. Late Mycenaean Warrior Tombs, in S. Deger-Jalkotzy and I. Lemos (eds.), *Ancient Greece. From the Mycenaean Palaces to the Age of Homer*, 150-179. Edinburgh, University Press.

Deger-Jalkotzy, S. 2008. Die Kriegervase von Mykene. Denkmal eines Zeitalters im Umbruch. In *Zeit der Helden. Die "Dunklen Jahrhunderte" Griechenlands 1200-700 v. Chr, Catalogue of the exhibition, Badisches Landesmuseum, Karlsruhe*, 76-83. Karlsruhe, Badisches Landesmuseum, Primus Verlag.

Dodds, E. (ed.) 1960. Euripides, *Bacchae*. Oxford, Clarendon.

Everson, T. 2004. *Warfare in Ancient Greece. Arms and armours from the heroes of Homer to Alexander the Great*. Stroud, A. Sutton Publishers.

German, C. S. 2005. *Performance, power and the art of the Aegean Bronze Age*. BAR International Series 1347. Oxford, BAR Publishing.

Graf, F. 2006. Initiation. A concept with a troubled history, in D. B. Dodd and C. A. Faraone (eds.), *Initiation in Ancient Greek Rituals and Narratives*, 3-24. London and New York, Routledge.

Guizzi, F. 1997. Terra comune, pascolo e contributo ai *syssitia* in Creta arcaica e classica. *Annali dell'Istituto Universitario Orientale di Napoli*, n.s., 4, 45-51.

Hall, J. M. 2007. *A History of the Archaic Greek World ca 1200-479 BCE*. Oxford, Blackwells.

Harrison, J. H. 1908-1909. The Kouretes and Zeus Kouros. A study in prehistoric sociology. *Annual of the British School at Athens* 15, 308-338.

Herzfeld, M. 1985. *The Poetics of Manhood. Context and Identity in a Cretan Mountain Village*. Princeton, University Press.

Iacovou, M. 1988. *The Pictorial Pottery of Eleventh Century BC Cyprus*. SIMA 78. Göteborg, Paul Åströms Förlag.

Iacovou, M. 1989. Society and settlements in Late Cypriot III. In. E. Peltenburg (ed.), *Early Society in Cyprus*, 52-59. Edinburgh, University Press.

Iles Johnston, S. 2006. 'Initiation' in myth; 'initiation' in practice: the Homeric Hymn to Hermes and its performative context, in D. B. Dodd and C. A. Faraone (eds.), *Initiation in Ancient Greek Rituals and Narratives*, 155-180. London and New York, Routledge.

Jeanmaire, H. 1939. *Couroi et Courètes. Essai sur l'éducation spartiate et sur les rites d'ado- lescence dans l'antiquité hellénique*. Lille, Bibliothèque universitaire.

Kaufman-Samaras, A. 1972. A propos d'une amphore Géometrique du Musée du Louvre. *Revue Archéologique* 1, 23-30.

Knapp, A. B. 2008. *Prehistoric and Protohistoric Cyprus: Identity, Insularity and Connectivity*. Oxford, University Press.

Langdon, S. 2008. *Art and Identity in Dark Age Greece, 110-700 B.C.E.* Cambridge, University Press.

Lebessi, A. 1991. Flagellation ou autoflagellation: Données iconographiques pour une tentative d'interpretation. *Bulletin de Correspondance Hellénique* 115, 99-123.

Lebessi, A. 1985. Το Ιερό του Ερμή και της Αφροδίτης στή Σύμη Βιάννου. I.1. Χάλκινα κρητικά τορεύματα (Βιβλιοθήκη της εν Αθήναις Αρχαιολογικής Εταιρείας 102). Athens, Archaeological Society.

Lebessi, A. 2002. Το Ιερό του Ερμή και της Αφροδίτης στή Σύμη Βιάννου. III: Τα χάλκινα ανθρωπόμορφα ειδώλια (Βιβλιοθήκη της εν Αθήναις Αρχαιολογικής Εταιρείας 225). Athens, Archaeological Society.

Marinatos, N. 2003. Striding across boundaries: Hermes and Aphrodite as gods of initiation, in D. B. Dodd and C. A. Faraone (eds.), *Initiation in Ancient Greek Rituals and Narratives*, 130-151. London and New York, Routledge.

Londsdale, S. H. 1993. *Dance and Ritual Play in Greek Religion*. Baltimore and London, The Johns Hopkins University Press.

Perlman, P. J. 1995. Invocatio and Imprecatio: the Hymn to the Greatest Kouros from Palaikastro and the oath in ancient Crete. *Journal of Hellenic Studies* 115, 161-167.

Price, S. 1999. *Religions of the Ancient Greeks*. Cambridge, University Press.

Rocchetti, L. 1994. *Sybrita. La valle di Amari tra Bronzo e Ferro* (IG 96). Rome, Gruppo Editoriale Internazionale.

Rombos, T. 1988. *The Iconography of Attic Late Geometric II Pottery*. SIMA pocket-book 68. Jonsered, Paul Åströms Förlag.

Rystedt, E. and Wells, B. (eds.) 2006. *Pictorial Pursuits: Figurative Painting on Mycenaean and Geometric Pottery*. Acta Instituti AthENiensis Regni Sueciae in 4°, 53. Stockholm, Svenska Institutet i Athen.

Sakellarakis, I. 1988. The Idean Cave: Minoan and Greek Worship. *Kernos* 1, 207-214.

Snodgrass, A. M. 1964. *Early Greek Armour and Weapons from the end of the Bronze Age to 600 B.C.* Edinburgh, University Press.

Sporn, K. 2002. *Heiligtümer und Kulte Kretas in klassischer und hellenistischer Zeit* (Studien zu antiken Heiligtümern 3). Heidelberg, Verlag Archäologie und Geschichte.

Spencer, P. 1985. Introduction: Interpretation of dance in anthropology. In P. Spencer (ed.), *Society and the Dance. The Social Anthropology of Process and Performance*, 1-46. Cambridge, University Press.

Stansbury-O'Donnell, M. 1999. *Pictorial Narrative in Ancient Greek Art*, Cambridge, University Press.

Stansbury-O'Donnell, M. 2006. The development of Geometric pictorial narrative as a discourse, in E. Rystedt and B. Wells (eds.), *Pictorial Pursuits: Figurative Painting on Mycenaean and Geometric Pottery*. Acta Instituti AthENiensis Regni Sueciae in 4°, 53, 247-253. Stockholm, Svenska Institutet i Athen.

Steel, L. 1993. The establishment of the city kingdoms in Iron Age Cyprus: an archaeological commentary. *Report Department of Antiquities Cyprus*, 147-156.

Stehle, E. 2000. Review of P. Ceccarelli, *La pirrica nell'antichità greco romana: Studi sulla danza armata*. Pisa and Rome, Istituti Editoriali e Poligrafici Internazionali, 1998, *Bryn Mawr Classical Review* 2000.03.17.

Tölle, R. 1964. *Frühgriechische Reigentänze*. Waldsassen, Bayern, Stiftland-Verlag.

van Wees, H. 1998. Greeks bearing arms: the state, the leisure class, and the display of weapons in Archaic Greece, in N. Fisher and H. van Wees (eds.), *Archaic Greece: New Approaches and New Evidence*, 333-378. London, Duckworth.

van der Vliet, E. Ch. L. 2011. The Early Greek Polis: Regime building, and the emergence of the state, in N. Terrenato and D. C. Haggis (eds.), *State Formation in Italy and Greece. Questioning the Neoevolutionist Paradigm*, 119-134. Oxford and Oakville, Oxbow Books.

Vermeule, E. Y. and Karageorghis, V. 1982. *Mycenaean Pictorial Vase Painting*. Harvard, University Press.

Wallace, S. 2006. The Gilded Cage? Settlement and socioeconomic change after 1200 BC: A comparison of Crete and other Aegean regions, in S. Deger-Jalkotzy and I. Lemos (eds.), *Ancient Greece. From the Mycenaean Palaces to the Age of Homer*, 619-664. Edinburgh, University Press.

Wallace, S. 2010. *Ancient Crete: From Successful Collapse to Democracy's Alternatives, Twelfth to Fifth Centuries BC*. Cambridge, University Press.

Weninger, B. and Jung, R. 2009. Absolute Chronology of the End of the Aegean Bronze Age, in S. Deger-Jalkotzy and A. Bächle (eds.), *LH IIIC Chronology and Synchronisms III. LH IIIC Late and the Transition to the Early Iron Age. Proceedings of the international workshop held at the Austrian Academy of Sciences at Vienna, February 23rd and 24th 2007*, 373-416. Vienna, Verlag der Österreichischen Akademie der Wissenschaften.

Whitley, J. 2000. *The Archaeology of Ancient Greece*. Cambridge, University Press.

Willets, R. F. 1955. *Aristocratic Society in Ancient Crete*. London, Routledge.

Willets, R.F. 1962. *Cretan Cults and Festivals*. London, Routledge.

Willets, R. F. 1965. *Ancient Crete. A Social History*. London, Routledge.

Yoffee N. 1997. The Obvious and the Chimerical. City-states in archaeological perspective, in D. L. Nichols and T.H. Charlton (eds.), *The Archaeology of City States. Cross Cultural Approaches*, 255-263. Washington and London, Smithsonian Press.

REVEL WITHOUT A CAUSE?
DANCE, PERFORMANCE AND GREEK VASE-PAINTING

Tyler Jo Smith

University of Virginia

Abstract*: Greek vase-painting provides one of the best sources for ancient Greek dance. Vases have long been used as evidence, but not always in the best or most responsible way. The online resources available today, such as the Beazley Archive Database and the Corpus Vasorum Antiquorum, allow scholars to approach dance iconography more efficiently and more systematically. This paper explains these resources and how best to use them, and summarizes how scholars of dance have made use of vase-painting over time. Using the earliest Athenian red-figure dance-scenes, mainly decorating drinking-cups, it is demonstrated how one category of visual and material evidence enables a certain type of archaeological analysis, and can ultimately broaden our knowledge of ancient Greek dance and performance in general.*

Ancient Greek dance has not always received the attention it deserves. The history of its scholarship was recounted in full up to 1997 by F.G. Naerebout, whose historiographical narrative nicely details key figures and approaches over a span of five centuries. His is the first attempt to summarize the little known twists and turns of a 'discipline' that belongs to the larger areas of Classics and Archaeology on the one hand, and histories of Art and Performance on the other. With such a dynamic and complex picture behind it, why have modern scholars not taken adequate notice? The reasons for this are perhaps many, and far beyond the scope of the present paper. Instead our concern here will be one particular category of visual and material evidence in an effort to demonstrate how a certain type of archaeological analysis can enlighten and broader our knowledge of ancient dance in general. It should become clear that Greek vases are an essential piece of the Greek dance enigma. In view of their quantity, portability, and of course imagery, there is much to be gleaned from them. What types of dances are portrayed on their surfaces? Who are the participants, and what is the setting? Do particular vase shapes lend themselves best to dance depictions?

Our first task will be to demonstrate a few tools of study for Greek vase-painting and how these can best be applied to the study of ancient Greek dance. How does one tackle the massive amount of information available without specialist knowledge of Greek connoisseurship and attribution, iconography and iconology? Secondly, we shall reverse the question, and ask how dance scholarship has made use of Greek vases, and why many of these methods have been problematic and should be read with caution. And finally, we shall offer as a case study the earliest representations of dance in Athenian red-figure vase-painting, and view these with specific iconographic criteria in mind. By isolating vases with dance scenes by date (*c.* 520-490), period (late Archaic), technique (red-figure), and, to an extent, form (drinking-cup), and thus employing the rigours of archaeological observation, it is thought that certain patterns of meaning will inform our knowledge of the bigger picture of ancient movement and performance.

Vases and Dance: Cups, Pots and Komasts

Greek vases representing scenes of dance occur in great numbers, in hundreds, perhaps even in the low thousands. The largest number of examples was made by Athenian craftsman working in the black- and red-figure techniques, active during the 6th and 5th centuries (i.e. the Archaic and Classical periods). That being said, scenes of dance on vases were also made in sizeable quantities in regional workshops outside Athens, especially in the black-figure technique (Smith 2010a), and the iconography of dance can itself be traced to the Late Geometric phase of vase-painting, when the earliest scenes of human-figure decoration occur (Boardman 1998, figs. 84, 89, 131). Interestingly, the Beazley Archive Database (BAD), a searchable online resource based at Oxford University and concerned primarily with Athenian vases, does not at present list any terms relevant to dance among its subjects under 'simple searching'; although under the category of 'advanced searching' one finds 'dance', 'dancer', and 'dancing', as well as '*komos*' (the ancient Greek word for a reveller or group of revellers). A search using the terms 'black-figure' and '*komos*' yields 898 records, while a search choosing the terms 'black-figure' and 'dancer' yields a mere six records (www.beazley.ox.ac.uk; 12/5/2009). A closer look at these results discloses the difficulty of using such data to generate even simple statistics or numbers. BAD no. 2411, a column-krater from Athens, contains the terms '*komos*' and 'dancing' in its description; as does BAD no. 697, a *kalathos* ('basket') attributed to the manner of the painter Elbows Out. For both examples, the reason for the overlap is easily explained. The full description of the first one reads: KOMOS, WOMEN DANCING; while the second is: KOMOS, YOUTHS AND WOMEN

IN CHITONISKOI DANCING, in each case indicating the general subject (*komos*) and its iconographic details (participants and action). To be certain, this is not the fault of the Beazley Archive, who work largely from published catalogues and other sources, and first and foremost from Sir John Beazley's 'lists': *Attic Red-Figure Vase-Painters* (*ARV*),1963, and *Attic Black-Figure Vase-Painters* (*ABV*), 1956 (Ashmole 1970; Smith 2005, 23).

Beazley himself, on the whole, preferred '*komos*' or 'komast' for his brief vase descriptions, the words largely retained today. He was not, however, averse to using other terms as demonstrated by a group of examples attributed to the Amasis Painter, each showing rather similar nude, male dancing figures in the black-figure technique:

1. 'Dionysos between two naked men dancing' (*ABV* 155.61).
2. 'Komos (male – fluting? – and man, between male and man dancing)' (*ABV* 155.66).
3. 'Dionysos with revellers (men and women)' (*ABV* 150.6).
4. 'Votaries of Dionysos' (*ABV* 151.11).
5. 'Woman handing lyre to a youth' (*ABV* 156.76).

In all fairness to Beazley, his interest was not in dance *per se*, regardless of descriptive terms; his was the much more daunting task of classifying all Athenian vases known to him by painter and group. However, the typology and terms for dancing figures or scenes on Greek vases have not been revised or reconsidered since the death of Beazley, although a number of publications pertaining to Greek dance, which make use of vases as evidence, have appeared since that time. Perhaps we should pause to ask ourselves: does the performer, his or her costume, accoutrements, venue, or companions indicate the need for a more nuanced terminology? Should '*komos*' (revel) or 'komast' (reveller) apply only to a rowdy, informal occasion, while 'dance' or 'dancer' denotes a more rhythmic and rehearsed one? A rigid binary relationship is not being proposed here, as visual analysis becomes even further complicated with the introduction of dance in other artistic media or forms such as bronze and terracotta statuettes. Despite potential pitfalls, Beazley's lists and the BAD must still be considered the starting point for vase research on dance related themes.

The second major resource available for vase-painting examples is the *Corpus Vasorum Antiquorum* (*CVA*). Founded in France in 1922, the *CVA* fascicules are publications of Greek vases in the museums of Europe, the United States and elsewhere (Rouet 2001, 124-37). In 2001 the Beazley Archive was awarded funding for the digitisation of the *CVA*, which it completed between 2002 and 2004. According to the website: "The project is on-going; new fascicules are being published and participating museums have the opportunity to contribute to the on-line database" (http://www.cvaonline.org/cva/ProjectPages/). As a result of this effort, the *CVA* may now be searched in tandem with the BAD, using either the 'simple search' option to locate an individual vase by country or museum, or by implementing an 'advanced search' across a wider range of fields (e.g. fabric, shape, technique, subject) and fascicules. The English language search terms used by the BAD are again found here, and a multi-language thesaurus has been supplied for shapes, painters and iconographic terms (thus also French, German, Italian, Spanish). An advanced search using exclusively the term '*komos*' (with thesaurus) yielded 2685 results, or 905 (without thesaurus) (www.cvaonline.org; 12/5/2009). As with the BAD search, this number must not be taken too seriously; a large percentage of these individual objects are cross-referenced with Beazley's lists, each has an assigned BAD number, and not all appear to list a *CVA* reference in their bibliography (perhaps because the BAD does not always include complete bibliography for each piece).

The publication of a recent *CVA* from the Archaeological Museum of Rhodes is a good example of how best to use *CVA* volumes for dance or other iconographic research, and a model of how authors of other volumes representing other countries or collections might be encouraged to present the same types of information (Lemos 2007). The author provides an 'index of principal subjects' (137-8), which includes 'komastss' [sic], and 'dancers' (under which she states simply 'see Komasts'). Under the heading of komasts she lists five vases (pls. 5.1, 16.2 and 4, 51.1-2, 52.1, 72.1-2). As to be expected, the photographs (multiple views of each object) are accompanied by descriptions, including information about technique, shape, condition, provenance, artist, date and iconography. Lemos takes full advantage of the space available to provide not only interpretation and comparanda, but also extensive subject bibliography. For example, Plate 16 is a black-figure amphora (inv. no. 15450) attributed to the Towry Whyte Painter, and dated *c*. 540-530 BC. On one side, three youthful, nude males take part in a running race, while two well-draped judges observe. The other side is described by Lemos as: "Komos with five naked youths with short hair dancing and pinching themselves and each other" (33). Such less-than-enthusiastic dancing figures are closely related to a number of contemporary examples made at the same time and in the same place (Smith 2010a, chapter 4). The two figures interpreted as 'pinching' are placed centrally in the scene and their actions are reminiscent of the 'up-and-down courtship' gesture often practised between male figures and in the presence of komast dancers on a large number of known examples (Beazley 1989, 4). Turning to the bibliography for this object, the author not only directs the reader to fundamental references pertaining to the figures on each side – running/athletics and dance/*komos* – but also to scholarship which connects the two, culturally and iconographically. This combined presentation again provides a vital starting point for further research, but, as with the BAD or Beazley's lists, one must be aware of the limitations of the evidence on offer. The *CVA*, if used properly, can be the best of all possible sources for illustrations and further reading, and we are very fortunate indeed that much of it is now within internet reach.

Dance and Vases: Then and Now

Early archaeological discoveries and their subsequent publication raised awareness of Greek dance imagery on vases and in other arts. To quote Naerebout:

> *"Decisive changes were brought about by the rapid development of classical archaeology. Of course objects carrying images of ancient Greek dance, primarily painted vases and sculptures, had been known for a long time, but they were few and mainly unpublished. Indeed, up to the 1750's authors on the dance hardly seem aware of the existence of any unwritten sources at all" (1997, 41).*

He goes on to mention the excavations of Pompeii and Herculaneum, and the beginnings of large-scale collecting and publication, which may be seen to mark the modern merging of Greek dance and Greek vases. This phase of vase scholarship, largely focused on the figure of Sir William Hamilton and others of Grand Tour fame, has been well documented and need not be repeated here (Nørskov 2002). Less inspiring for our purposes are the next 100 years or so. Let us fast forward to the end of the 19th century, the beginning of the next important moment for dance and vases (Naerebout 1997, 54-71), when we encounter the first of a series of publications on Greek dance that can be isolated for its treatment of Greek vases as evidence.

The publication of Maurice Emmanuel's *La Danse grecque antique d'après les monuments figurés* in 1896 is the first book on the subject to make abundant use of Greek vases. Emmanuel believes vase-painting to be among the most important subjects for Greek dance history, even privileging it above sculpture, and summarizing early in the opening chapter the styles, techniques, and basic use of dance themes (1984, 7-13). His assumption that the figures on vases might be used to reconstruct the gestures, positions, and steps of ancient Greek dance is difficult to accept today, and created a certain amount of controversy at the time, as in the following years (Naerebout 1997, 61-2). Emmanuel's literal reading of vase imagery is further complicated by his attempt to relate the figures and forms to contemporary French ballet. It is not unusual in the book to find line drawings of modern dance steps interspersed with those revealing the inspiring details of a black- or red-figure vase (**figs. 1 and 2**). Such is the case where he shows the figures of several dancing satyrs from the same red-figure vase in relation to a series of known modern dance positions (1984, 184, figs. 286-398), claiming: "Les trios Satyres à pieds de bouc, de la fig. 386, sont empruntés à un même vase. Malgré la déformation de leurs membres inférieurs, ils sont facilement comparables aux images de la série chronophotographie qui commence à la figure 387 et finit à la figure 398». The author also uses contemporary chronophotography (a combined study of movement and art) to illustrate such a series in action; and thus a woman dressed in 'Greek' garments, à la Isadora Duncan, executes the steps that are again found juxtaposed with elementary

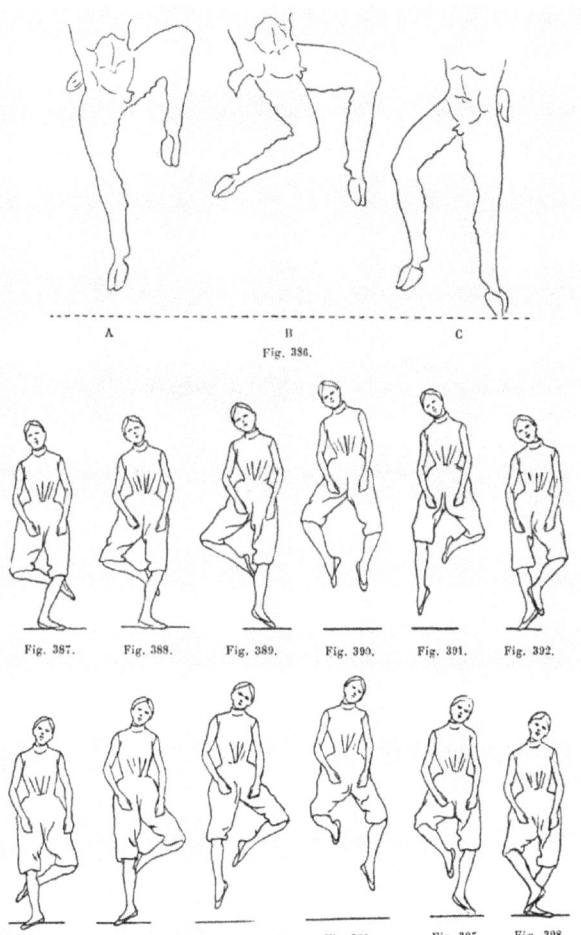

Figure 1. Legs and feet of three dancing satyrs from an ancient vase and a chronophotographic series demonstrating similar steps (after Emmanuel 1984 [1896], 184-5).

drawings derived from the figures on a single vase, or indeed from several. Their movements are described as: « il Dégage et Plie, - il Saute, - il retombe en Jetant» (Emmanuel 1984, 212, no. 320; cf. Séchan 1930, 202, fig. 46).

The next book-length publication of importance to our present topic is Louis Séchan's *La Danse grecque antique* of 1930. Like Emmanuel before him, Séchan's engagement with Greek vases and their images, and indeed with Greek art in general, is largely for the sake of reconstructionism – i.e. bringing to life the dances of the ancient past, both named and unnamed. To this end, he makes extensive reference to ancient textual sources, such as Plato's *Laws*, for the many "variétiés des danses" (Séchan 1930, 57-83), be they sacred or secular, public or private. Indeed, a large percentage of the book is "devoted to modern revivalists" (Naerebout 1997, 74), among them Isadora Duncan herself (Naerebout 1997, 62-66; Smith 2010b). Another similarity between the two Frenchmen, Emmanuel and Séchan, is the shared misunderstanding of Greek vase-painting as an art form all its own, subject to everything from technique and convention, to market and artistic whim. Séchan's book was reviewed in *Classical Review* (1931) by none other than Sir

Figure 2. Dancing maenads from an ancient vase and modern dancer similarly posed (after Emmanuel 1984 [1896], 202-3).

John Beazley himself, who, not surprisingly, criticized the author's comprehension and use of iconography. However, the review by E.N. Gardiner in the *Journal of Hellenic Studies* in 1930, while less critical than Beazley's, sums up well both the intentions and mentality of the author:

> *"From an examination of the positions and movements represented on vases, the writer concludes that the Greek dance exercised every part of the body, and in particular that there was far greater freedom in the use of the arms and hands than in the orthodox modern ballet. But he is less concerned with the details of movement than with the spirit of the dance"* (350).

The reconstructionist trend continues with G. Prudhommeau, *La Danse grecque antique* (1965) about which all of the above would apply, and then some (Naerebout 1997, 86-8).

Another scholar of Greek dance, who published almost obsessively on the topic, was Lillian B. Lawler. Her book, entitled *The Dance in Ancient Greece* (1964a), is still considered the seminal and, in fact, only introductory work in English (Naerebout 1997, 81). In the spirit of Séchan, Lawler uses similar (but not exactly the same) categories (orgiastic, festivals, etc.) to frame her presentation of dance.

She also draws heavily on the literary corpus for both lead and support. Her haphazard use of Greek vases has been discussed elsewhere (Smith 2010b; Ley 2003, 477-8), but suffice it to say that her visual comprehension is lacking in sophistication.

A second book by Lawler, *The Dance of the Ancient Greek Theatre* (1964b), should be mentioned as part of a larger group of publications focused exclusively on dance in relation to theatre, both dramatic origins and performance. Classical scholars in Britain such as A. D. Pickard-Cambridge and T. B. L. Webster devoted a good deal of scholarly attention to this area. Webster, for example, published *The Greek Chorus* in 1970 with little fanfare (Naerebout 1997, 89, n. 261). In the book he attempts to unearth the evidence for choral performance, drawing on the evidence of iconography and text. Better known, but certainly problematic, was his joint venture with A.D. Trendall, *Illustrations of Greek Drama* (1971), where sincere efforts were made to match the images on Greek vases with both surviving and lost plays. Their catalogue includes some dancing figures, namely komasts or choruses, who they see as belonging to the category of "pre-dramatic monuments" (15-27). Similarly, Ghiron-Bistagne (1976) places her discussion of '*komos*' in an appendix subtitled "recherches sur l'origine des genres scéniques" (207-96). Related to these studies of dance via drama, we might also list the more specialized studies of Archaic komasts or 'padded dancers', starting from Greifenhagen's 1929 *Eine attische schwarzfigurige Vasengattung und die Darstellung des Komos im VI. Jahrhundert* up to the most recent (Smith 2010a), not to mention the large number concerned with Dionysian dance themes, choruses and dithyramb (Naerebout 1997, 122-31).

M. H. Delevaud-Roux's three volumes on various aspects of Greek dance, among them pyrrhic (1993), 'pacifique' (1994), and Dionysian (1995), continues the tradition of her French predecessors (1993, 10-15). She downplays the ballet, however, embracing other traditions of dance, among them the folk-dances of Greece and other cultures (1993, 15-16, 18-20). The research itself is highly speculative and is in no way a literal treatment of the evidence. Her understanding of vase iconography, on which she relies heavily, is basic. She makes good use of both Beazley and the *CVA*, but illustrates her books with simple sketches or renderings, as opposed to the high quality photographs or professional drawings we have come to expect of such publications. Her reverence for and reliance on Emmanuel a century on is truly astounding. The greatest merit of her approach is its grounding in professional performance, and also because she conveniently collects in one place many representative examples of each of her three thematic areas.

What is the moral of this short fable about dance and vases? To be fair, the very select list of publications presented above comprises a tiny percentage of the whole of Greek dance scholarship that has incorporated Greek vases. These examples have been chosen for their influence and longevity. Thus, what is the current direction of dance scholarship in this regard? It seems the latest trend is twofold, and is very much in keeping with some of Classical scholarship's broader aims: performance and reception. Performance studies have become more a staple of publications and conferences in the Classics than they were a generation ago, and some have even combined with reception (Revermann and Wilson 2008; Hall and Wyles 2008). As we have seen, many Greek dance scholars who incorporate vases into their program of research are themselves performers, but none to my knowledge has embraced performance as a theoretical category. Classical scholars are finally beginning to look beyond the standard performance locales of the past, such as the theatre itself. Rather, in an effort to gain a clearer understanding of dress, ritual, festival, and public or private spectacles of various types, it is now possible to speak in terms of performance spaces and events more widely understood (e.g. Naerebout 2006; cf. Raftis 1987, 21). Indeed the 'performing arts' – dance, drama, music – are only the starting point of this discussion. Meanwhile, reception studies have taken hold in Classical scholarship in a big way, and dance has now formally entered the corpus with the publication of *The Ancient Dancer in the Modern World* (Macintosh 2010). One paper therein is devoted to the place of vases in the study of Greek dance reception, with particular attention to Isadora Duncan and Lillian Lawler, and their individual, if rather different contributions (Smith 2010b; cf. Ley 2003). It is undoubtedly safe to say that the time has come for a new book on ancient Greek dance that makes full and credible use of art, archaeology and iconography on the one hand, and the corpus of literary, historical and philosophical texts on the other (Raftis 1987, 25).

The First Red-Figure Scenes: Revels and Causes

> "The rude beginnings of Greek tragedy can be observed in the clownish performances which are represented in all but unlimited numbers on the black-figured vases, and on those of the red-figured style during the first twenty-five years or so. Such dances as one may note on every hand in a vase collection, furnish the keynote to the dithyrambic movements that ushered in the tragic drama of Thespis and his successors, and nearly all belong to a time when there was no libretto and no prescribed acting. Complete absence of dignity, we may be sure, characterized every feature of these dances" (Huddilston, 91).

The above quote, written in 1902, demonstrates how much our thinking has matured in regard to dance iconography on vases. The author over-generalizes the situation, revealing an incomplete knowledge of dance and early drama, not to mention a rather puritanical take on the images themselves. He is surely referring to the *komos* and komasts produced in Athens in both black-and red-figure, and mentioned in the first section of this paper. But what exactly did Huddilston observe on the surfaces of these vases? Which ancient versions is he lumping together here? It seems fairly obvious that the dances with their "complete absence of dignity" are *both* those black-figure examples with male dancers, either in 'padded' clothing or fully nude, who slap their bottoms and engage playfully with friends (Smith 2010a), *and* the red-figure cups with nude male revellers, lightly adorned with boots, cloaks and walking sticks en route to or from a *symposion* (drinking party) (Boardman 1975, 219; Vierneisel and Kaeser 1990, 283-98). Although both groups have been classified by modern scholars by the terms '*komos*' and 'komast', the scenes, locations and participants are often different. It is the second group that claims our attention here.

In *The World of Greek Dance* (1987), A. Raftis examines the subject across time and place as both a 'cultural phenomenon' and a social one (21-4). He begins his section on 'Dance in Ancient Greece' with a stab at modern scholarship for its "quote an excerpt from an ancient text or include a representation from a Classical vase" approach (25). He appreciates the 'disjointed' nature of the material and the difficulties of interpretation, regardless of whichever type of evidence is employed. With regard to Greek folk-dance, he divides the discussion into two main categories: 'dance situations' and 'elements of traditional dance' (38-51, 52-77). The former refers to the venues and occasions - from weddings and family celebrations, to the *kafeneion*, Carnival and Easter – to be certain each constitutes a modern version of an ancient setting. His latter 'elements' category is also applicable to the ancient world up to a point: the costumes, music, and instruments associated with particular dances and regions. As we look closely at Greek vases, regardless of technique, date, or place of manufacture, it is sensible to apply similar criteria of setting and appearance. Vase-painters regularly apply specific visual clues, if minimal accuracy, to their portrayals of the three-dimensional art of dance on the two-dimensional surface of the vase.

Let us turn our attention now, as promised, to the earliest scenes of dance appearing on Athenian vases produced in the red-figure technique. It is the first practitioners of a full red-figure technique, known as the 'Pioneers' or 'Pioneer Group', active *c.* 500 BC, who "seized upon the essence of what red-figure could offer, refining it in detail of drawing and composition" (Boardman 1975, 29). These painters were especially interested in the human (i.e. male) body, resulting in visual experiments in athleticism and movement. They represent a critical moment of transition between the austerity of the Archaic style and the evolving fluidity of the Classical (Stewart 2008, 12-17). It is worth stressing that the black-figure technique, which dominated vase decoration from the beginning of the 6th century BC, created hundreds of scenes of dance on cups and larger shapes. That being said, black-figure painters never attempted the amount of movement or detail in dance that they later would once red-figure had been invented (Smith 2003). As we confront this change of technique from black-to red-figure at *c.* 520 BC, and its application to images of dance, we are forced to question the possibility of a change in meaning. Do new dancers and styles emerge concurrent with this artistic shift?

The obvious starting point of discussion is the amphora signed by Euthymides with three dancing, nude males decorating one side (**fig. 3**) (Munich 2307; Boardman 1975, fig. 33.2). The composition is deliberately well-balanced, as the two dancers on the ends raise a knee towards the slightly more stationary dancer at the centre. The gestures of the figures are varied, as are their body positions. A quick comparison with other works by this same artist reveals his interest in painting the nude male body from various angles, including full back and 3/4 views (Boardman 1975, figs. 36-7). What defines this particular image as a dance scene, however, is more than the poses and gestures of the performers. Their attributes and clothing, such as their cloaks, and the stick held by one, associate them with many contemporary dance scenes, and indicate their mobility: they are 'dressed' to go out; while their *hypothymides* (party wreaths) and the fancy *kantharos* (high-handled drinking-cup) held by one, recall a number of black-figure dance scenes from the previous decades, and lend a Dionysian air to the occasion. The most notable features of the scene, however, are the inscriptions, which name the artist and also the figures. The one to the far left, who holds the *kantharos* is conveniently labeled *komarchos* or 'leader of the revel' (Robertson 1992, 33; Hoppin 1917, 12).

Another Pioneer Group artist, known as the Dikaios Painter, incorporates some of the same elements, though in a slightly different manner, on a *psykter* (wine-cooler) in the British Museum (**fig. 4**) (E 767; Boardman 1975, fig. 47). One side is identified by Beazley as a '*komos*', and again it is the combination of poses and adornment that confirms a dance scene (*ARV* 31.6). Two balding males with long beards perform behind a diminutive nude and youthful figure. The larger dancing males are draped in cloaks and boots, one carries a stick and stemmed drinking-cup, and the other plays the lyre to accompany their movements. The lyre-player, although once identified as Dionysos, is described by Beazley in *ARV* as a "poet" (cf. Hoppin 1917, 64-8). The two men are looking in opposite directions, yet each moves to the right. The younger boy also dances right, and seems largely unaware of the older men who follow him. The two male figures on the opposite side, playing flutes, are no doubt related to the others and attend the same event.

The largest number of komast scenes in early Athenian red-figure actually decorate cups, as opposed to the larger 'pots' we have just mentioned. In fact, it has been observed that "eight out of ten of the red figure vases which survive from the first generation in the new technique are cups" (Boardman 1975, 55). This may explain the prevalence for dance iconography on them during this period, and at least two points should be made. Firstly, the dancers (or revellers or komasts or whatever we decide to call them) are attending a *symposion*, the men's-only drinking party well-attested in ancient literature (Lissarrague 1990, 8-11). The painters, however, heavily abbreviate the event, whittling it down to enough basic elements to make it instantly recognizable. Secondly, the shapes themselves, both 'pots' and cups, are suitable for use at a *symposion*, where drinking was the central activity. Thus we get a sense of participants (men and boys), their activities (dancing and drinking), and their mobility (cloaks, sticks, boots). What is not always clear is whether the figures are arriving, departing, or in the midst of the party. The tradition of painting dance scenes on cups, including some with *symposion* settings, was well-established in Athenian black-figure (Smith 2000; Vierneisel and Kaeser 1990, 222-7). Although the red-figure cup-painters seem simply to carry on the tradition, the appearance and attributes of the dancing figures is not always the same as the earlier examples and technique. Indeed, by the mid 6th century, most dancing figures on black-figure cups and other shapes are full nudes. Red-figure painters prefer to cloth the dancing male body, if only scantily (Stewart 2008, 14). Interestingly, once the red-figure technique is established, the black-figure painters embrace the new red-figure fashion for dancers: cloak, sticks, boots, etc. Such is the case with the mature male represented on the interior of one black-figure cup in Paris attributed to the Haimon Painter (Louvre CA 3110; *ABV* 561.535), who has placed his stick and cloak on the ground in order to dance more energetically with both hands raised.

Cup-painters, regardless of time or technique, were limited by the odd, unusual and limited space available for decoration. As a result, those working in the early red-figure technique create clever solutions to the merging of shape and image, without losing sight of their new-found love of motion and emotion. Painters, such as Oltos and Epiktetos, and their contemporaries such as Skythes and the Epeleios Painter, are especially fond of our subject (*ARV* chapters 4-6). In some instances, the interior of the cup, known as the 'tondo', shows a dancer in black-figure, while on the exterior are related scenes in red-figure, such as one ascribed to Oltos and now in Basel (Antikenmuseum und Sammlung Ludwig Lu33; *ARV* 44.79). The term commonly used for vases which combine the two techniques is 'bilingual' (Robertson 1992, 9-14). In this particular example, it is notable that the black-figure dancer on the inside is wearing the cloak, wreath and boots of his red-figure companions. The number of red-figure (or 'bilingual') examples relevant to our topic and dated to this phase is far too great to discuss in any detail, but here we shall highlight a few that may stand for the many.

The overall party atmosphere of the *symposion* is invoked on the many cups that isolate an individual dancer or participant. At times, it is difficult at a glance to distinguish running from dancing, but again attributes come to our aid. The beardless and fully nude male, who moves right and looks left (a standard composition) on the tondo of a cup in Oxford is one such figure (Ashmolean Museum 516; Boardman 1975, fig. 60). However, his drinking-horn and wreath, coupled with the *symposion* on the exterior of the same object, clarify the situation (*ARV* 63.92). On another

Figure 3. Detail of Athenian red-figure amphora with three dancers, Euthymides (Munich, Staatliche Antikensammlungen 2307; after Hoppin 1917, pl. 1 [E I]).

*Figure 4. Details of Athenian red-figure **psykter** with dancers, Dikaios Painter (London, British Museum E 767; after Hoppin 1917, pl. XVIII-XIX [E 7]).*

example a cloaked, wreathed, and otherwise naked male revels to the right, armed with castanets in each hand and the cover for his double-pipes draped over one arm, reminding us that dance, no matter how casual, is accompanied by music or percussion (Cambridge, Fitzwilliam Museum 49.1864; *ARV* 111.14). A single male komast expectedly bedecked with cloak, wreath and walking stick, and positioned with both knees bent, favours drink to music; he awkwardly balances a large stem-less cup in the palm of his hand, and his pipes case is placed beside him (**fig. 5**) (*ARV* 178.5 [once London, Mitchell]; Boardman 1975, fig. 125). The Greek word *'kalos'* is part of the standard inscription naming a boy (in this case 'Akestor') as "beautiful" (Smith 1896); but this dancer almost appears to sing the sentiment directly from his mouth.

A fondness for party-tricks, such as balancing a cup or an amphora (e.g. Munich, Staatliche Antikensammlungen 8709; *ARV* 82.2; Boardman 1975, fig. 68; and Paris, Louvre F 129; *ARV* 84.20) or straddling a wineskin (Paris, Louvre G70, *ARV* 169.6) are further manifestations of all that is possible in the space of the *symposion* or the imagination of the painter (Lissarrague 1990, 47-67). In each example, regardless of placement, the focus is on the individual activity of the individual reveller, and the painters are exceptionally versatile in their application of iconographic details. Small group compositions are also possible. The cup tondo in the British Museum (**fig. 6**) (1843.1103.9 [E 38]) depicting a semi-nude girl who plays castanets and dances draped in an animal-skin and turban to the tunes of cloak-clad piper, has been identified as a *'komos'* (*ARV* 72.15; Boardman 1975, fig. 75.2). Her exotic clothing, however, beckons other possibilities. While she may be a hired performer entertaining at a party, her overall dress and appearance categorize her as 'other'. With that idea in mind, we may now ponder the exterior of the same vessel, which embraces on each of its two sides these multiple levels of meaning. On one we find the usual *symposion*, but on the other a portrayal of Herakles and Busiris – a myth that exemplifies the dynamic between Greeks and 'others' (cf. Lissarrague 1990, 11, 91-3; Todisco 2006, 131-55).

Two red-figure vases, each attributed to Epiktetos, verify the importance of iconography and occasion, as well further artistic possibilities. The first is a cup in the Ashmolean Museum (Oxford 520; Boardman 1975, fig. 76) featuring on one side two lively nude male dancers, who Beazley labels as 'cup-bearers' (*ARV* 76.84). Interestingly, these revelling males, who handle drinking vessels of various shapes, are otherwise lacking the standard attributes and dress we have come to associate with first generation red-figure depictions. Rather, the painter gives pride of place to the most powerful element of *komos* and *symposion* from the earlier black-figure tradition: the *krater*. The shape is attested in ancient sources as the mixing-bowl used at a *symposion* for mixing and serving wine, and thus functioning as a sort of party punch-bowl (Vierneisel and Kaeser 1990, 299-302). Although many Athenian or other black-figure vase-painters incorporated the *krater* in scenes of komast dancing, in the

Figure 5. Athenian red-figure cup with komast, the Salting Painter (Once London, Mitchell; after Smith 1896, pl. 13).

Figure 6. Athenian red-figure cup with dancing girl and piper, Epiktetos (London, British Museum 1843.1103.9 [E 38] © The Trustees of the British Museum).

red-figure examples discussed above, the shape was not the focus. Instead the painterly emphasis was on the individual, the dance, and the body, and the *symposion* was implied in other ways. In this version, by contrast, Epiktetos smacks us over the head with sympotic symbolism, and also shares the importance of dance and performance in relation to that occasion.

One last example again verifies the varied and eclectic nature of early red-figure iconography, but reminds us of our common theme of dance. The shape chosen for decoration is a plate (London, British Museum E 137; Boardman 1975, fig. 78; *ARV* 78.95), presenting the painter with more or less the same 'canvas' as the circular cup tondo. The scene of two males, one bearded, the other beardless, whose nude figures overlap, is aesthetically pleasing and subtle. The younger, beardless male stands erect and plays the pipes, while his older, bearded companion, dressed in a cloak and boots, bends towards the floor to pick up a large open vessel. Again, the setting, though not overtly stated, must be a *symposion*. At the same time, a plethora of details are available to the viewer: age, (homo)sexuality, drink, movement, music. Here the dancing has paused. We are forced to meditate more quietly on this snapshot of reality, where *komos* and *symposion*, dancing and drinking, unbridled entertainment and light-hearted fun, once again converge.

As modern viewers of vases and enthusiasts of dance, we may only imagine the space of the *symposion*, with its drinking, dancing and discussion. Those rowdy revellers who crash the party at the moment of ultimate intoxication encouraged their aristocratic audience to put down their drinks temporarily, take the stage, and lose control. As this very limited survey has demonstrated, the issues and complications of uniting dance and vase scholarship are manifold. The unique history of each area examined through modern eyes has not always made the most of the evidence or examples on offer. With much of the ground-work laid by previous generations of dance and vase scholars, the virtues of performance theory and experimental archaeology might now be questioned more seriously. The 'situations' and 'elements' of both ancient Greek art forms – dance and vase-painting – reveal that the potential of these two areas combined has yet to be realized.

Acknowledgments

Thanks are owed to Kathryn Soar and Christina Aamodt for inviting my contribution to the volume, despite not being able to attend the conference. Much thinking and research pertaining to this paper occurred at the archives of the Greek Dances Theatre 'Dora Stratou' in Athens during spring 2008; and at the Archive of Performances of Greek and Roman Drama at Oxford University, where I was Visiting Research Fellow during summer 2009. My appreciation is extended to both Alkis Raftis and Fiona Macintosh for their hospitality and inspiration. The title was inspired by C. Segal, *The Death of Comedy* (Cambridge: Harvard University Press, 2003), 7. The illustrations were prepared by Dan Weiss, to whom I am truly grateful.

Abbreviations

ABV J. D. Beazley, *Attic Black-Figure Vase-Painters* (Oxford, 1956).
ARV J. D. Beazley, *Attic Red-Figure Vase-Painters*, 2nd edn. (Oxford, 1963).
BAD Beazley Archive Database
CVA Corpus Vasorum Antiquorum

References

Ashmole, B. 1970. Sir John Beazley. *Proceedings of the British Academy* 56, 443-61.
Beazley, J. D. 1956. *Attic Black-Figure Vase-Painters*. Oxford, Oxford University.
Beazley, J. D. 1963. *Attic Red-Figure Vase-Painters*, 2nd edn. Oxford, Oxford University.
Beazley, J. D. 1989. *Some Attic Vases in the Cyprus Museum*, ed. by D. C. Kurtz. Oxford, Oxford University.
Boardman, J. 1975. *Athenian Red Figure Vases: the Archaic Period*. London, Thames and Hudson.
Boardman, J. 1998. *Early Greek Vase Painting*. London, Thames and Hudson.
Delavaud-Roux, M.-H., 1993. *Les Danses armées en Grèce antique*. Aix-en-Provence, Université de Provence.
Delavaud-Roux, M.-H. 1994. *Les Danses pacifiques en Grèce antique*. Aix-en-Provence, Université de Provence.
Delavaud-Roux, M.-H. 1995. *Les Danses dionysiaques en Grèce antique*. Aix-en-Provence, Université de Provence.
Emmanuel, M. 1984. *La Danse grecque antique d'après les monuments figurés*. Reprint of 1896 edn. Paris/Geneva, Slatkine.
Ghiron-Bistagne, P. 1976. *Recherches sur les acteurs dans la Grèce antique*. Paris, Les Belles Lettres.
Greifenhagen, A. 1929. *Eine attische schwarzfigurige Vasengattung und die Darstellung des Komos im VI. Jahrhundert*. Königsberg, Gräfe und Unzer.
Hall, E. and Wyles, R. (eds) 2008. *New Directions in Ancient Pantomime*. Oxford, Oxford University.
Hoppin, J. C. 1917. *Euthymides and his Fellows*. Cambridge, Mass., Harvard.
Huddilston, J. H. 1902. *Lessons from Greek Pottery*. London, Macmillan.
Lawler, L. 1964a. *The Dance in Ancient Greece*. London, A. & C. Black.
Lawler, L. 1964b. *The Dance of the Ancient Greek Theatre*. Iowa City, University of Iowa.
Lemos, A. A. 2007. *Corpus Vasorum Antiquorum. Rhodes, Archaeological Museum: Attic Black Figure* 1, Greece no. 10. Athens, Academy of Athens.
Ley, G. 2003. Modern Visions of Greek Tragic Dancing. *Theatre Journal* 55, 467-80.
Lissarrague, F. 1990. *The Aesthetics of the Greek Banquet: Images of Wine and Ritual*, trans. by A. Szegedy-Maszak. Princeton, Princeton University.

Macintosh, F. (ed.) 2010. *The Ancient Dancer in the Modern World: Responses to Greek and Roman Drama*. Oxford, Oxford University.

Naerebout, F. G. 1997. *Attractive Performances. Ancient Greek Dance: Three Preliminary Studies*. Amsterdam, J. C. Geiben.

Naerebout, F. G. 2006. Moving Events: Dance at Public Events in the Ancient Greek World. Thinking through its Implications. In E. Stavrianopoulou (ed.), *Ritual and Communication in the Graeco-Roman World*, 37-67. Liège, Centre international d'étude de la religion grecque antique.

Nørskov, V. 2002. *Greek Vases in New Contexts*. Aarhus, Aarhus University.

Prudhommeau, G. 1965. *La Danse grecque antique*. Paris, Centre national de la recherche scientifique.

Raftis, A. 1987. *The World of Greek Dance*. Athens, Finedawn.

Revermann, M. and Wilson, P. (eds) 2008. *Performance, Iconography, Reception: Studies in Honour of Oliver Taplin*. Oxford, Oxford University.

Robertson, M. 1992. *The Art of Vase-Painting in Classical Athens*. Cambridge, Cambridge University.

Rouet, P. 2001. *Approaches to the Study of Attic Vases. Beazley and Pottier*. Oxford, Oxford University.

Séchan, L. 1930. *La Danse grecque antique*. Paris, E. De Boccard.

Smith, C. 1896. A Kylix with a New Kalos Name. *Journal of Hellenic Studies* 16, 285-7.

Smith, T. J. 2000. Dancing Spaces and Dining Places: Archaic Komasts at the Symposion. In G. R. Tsetskhladze, A. J. N. W. Prag and A. M. Snodgrass (eds), *Periplous: Papers on Classical Art and Archaeology Presented to Sir John Boardman*, 309-19. London, Thames and Hudson.

Smith, T.J. 2003. Black-Figure Komasts in 'the Age of Red-Figure': Continuity or Change? In B. Schmaltz and M. Söldner (eds), *Griechische Keramik im kulturellen Kontext. Akten des Internationalen Vasen-Symposions in Kiel vom 24. bis 28.9.2001 veranstaltet durch das Archäologische Institut der Christian-Albrechts-Universität zu Kiel*, 102-4. Münster, Scriptorium.

Smith, T. J. 2005. The Beazley Archive: Inside and Out. *Art Documentation* 24, 22-5.

Smith, T. J. 2010a. *Komast Dancers in Archaic Greek Art*. Oxford, Oxford University.

Smith, T. J. 2010b. Reception or Deception? Approaching Greek Dance through Vase-Painting. In F. Macintosh, F (ed.), *The Ancient Dancer in the Modern World: Responses to Greek and Roman Drama*, 77-98. Oxford, Oxford University.

Stewart, A. 2008. *Classical Greece and the Birth of Western Art*. Cambridge, Cambridge University.

Todisco, L. 2006. *Pittura e ceramica figurate tra Grecia, Magna Grecia e Sicilia*. Bari/Rome, la Biblioteca.

Trendall, A. D. and Webster, T. B. L. 1971. *Illustrations of Greek Drama*. London, Phaidon.

Vierneisel, K. and Kaeser, B. (eds) 1990. *Kunst der Schale: Kultur des Trinkens*. Munich, Staatliche Antikensammlungen und Glyptothek.

Webster, T. B. L. 1970. *The Greek Chorus*. London, Methuen.

www.ingramcontent.com/pod-product-compliance
Lightning Source LLC
Chambersburg PA
CBHW041709290426
44108CB00027B/2906